APR '98

ENVIRONMENTAL HAZARDS

TOXIC WASTE AND
HAZARDOUS MATERIAL

ENVIRONMENTAL HAZARDS

TOXIC WASTE AND HAZARDOUS MATERIAL

A Reference Handbook

E. Willard Miller
Department of Geography

Ruby M. Miller
Pattee Library

The Pennsylvania State University

CONTEMPORARY WORLD ISSUES

ABC-CLIO

Santa Barbara, California
Denver, Colorado
Oxford, England

Library of Congress Cataloging-in-Publication Data

Miller, E. Willard (Eugene Willard), 1915–
    Environmental hazards : toxic waste and hazardous material : a reference handbook / E. Willard Miller, Ruby M. Miller.
        p. cm. — (Contemporary world issues)
    Includes bibliographical references and index.
    1. Hazardous wastes. 2. Hazardous waste sites. 3. Pollution.
    4. Environmental law. I. Miller, Ruby M. II. Title. III. Series.
    TD1030.M56 1991   363.72'87—dc20   91-14545

ISBN 0-87436-596-1 (alk. paper)

98 97 96 95 94 93   10 9 8 7 6 5 4 3 2

ABC-CLIO, Inc.
130 Cremona Drive, P.O. Box 1911
Santa Barbara, California 93116-1911

Contents

4 Directory of Organizations, 129

5 Bibliography, 149

Preface

THE BENEFITS OF AN INDUSTRIAL SOCIETY have made it possible for people to ignore for a long time the damage being inflicted upon the environment. By the 1960s, the widespread use of pesticides created a fear for public health and the land, which was also being polluted as a result of the disposal of hazardous chemicals and biological by-products in streams and at dump sites.

By the 1970s, it was evident that the pollution of the land was a national problem and required the development of national controls. The Environmental Protection Agency found that the scope of the damage done by hazardous waste sites and the pollution of the environment was far greater than had originally been thought. It is now recognized that the cleanup of the environment will continue for many years and will be very costly.

This volume begins with an introduction that includes a discussion of the definition and generators of wastes and a description and analysis of the evolution of the awareness of toxic waste and hazardous material in the environment. In the 1960s, Rachel Carson's superb *Silent Spring* launched a massive attack on the use of pesticides. This resulted in the first effective legislation to control pesticides in the environment. These efforts were followed by concern for the control of toxic and hazardous chemicals created by our technological society and the establishment of a national program focusing on the cleanup of hazardous waste sites. The introductory chapter concludes with a discussion of some environmental toxic chemicals, waste control technology developments, the transportation of hazardous wastes, and a review of petroleum environmental pollution and asbestos control.

The chronology in Chapter 2 lists some of the critical dates for pesticide control, major toxic waste sites, chemical accidents, major oil spills, and laws and regulations.

The fundamental laws to control environmental degradation are the Resource Conservation and Recovery Act of 1976 and the Comprehensive Environmental Response, Compensation and Liability Act of 1980 (Superfund). These laws, establishing the nation's basic hazardous material management system, provided the authority to encourage the conservation and recovery of valuable materials. Other laws concern specific aspects of toxic waste management, such as control of asbestos, regulation of lead contamination, and pesticide monitoring. Chapter 3 provides an overview of these and other key laws and legislation relating to toxic waste and hazardous material.

Many organizations have been established to consider environmental problems. Chapter 4 provides a directory of these organizations, divided into four major parts: U.S. government agencies, U.S. intergovernmental advisory committees, private organizations in the United States, and international organizations.

The available literature on environmental problems has grown greatly. Many viewpoints are now available as to how environmental problems can be solved, and the literature varies from scientific to socially oriented studies. Chapter 5 lists a number of recently published reference materials, books, journal articles, and government documents, as well as the leading journal titles, and Chapter 6 lists the nonprint materials available on the subject. The volume concludes with a short glossary and an index.

E. Willard Miller
Ruby M. Miller
The Pennsylvania State University

1

Toxic Waste and Hazardous Material: A Perspective

THE MODERN ENVIRONMENTAL MOVEMENT has its roots in the distant past. As early as 300 B.C., Theophrastus described plant diseases known in modern times as rust, scab, rot, and scorch. Sulfur and arsenic traditionally have been the chemicals used to control pests in the environment. In the nineteenth century, the ill effects of a contaminated environment on health began to be recognized. In 1849, for example, Dr. John Snow in London noted that an outbreak of cholera was not evenly distributed in the city, but was greatest in areas where water had been contaminated by sewage flowing from a particular pump.

Beginning in the twentieth century, the preservation and protection of the environment became a theme of many scientists and writers. In 1913, Herbert Zurich asserted in his book *The Good Ship Earth* that life on earth depended on a closed system of living things and inanimate objects. In 1927, Edward A. Ross was among the first to write on modern population growth. At that time many herbicides and insecticides were developed to control pests in order to increase the food supply of the world. The control of insects was of paramount importance; little thought was given to possible contamination of the environment in the process.

The modern environmental movement can be traced to the wide spread fear of the effects of radioactive contamination of the environment due to the atmospheric testing of nuclear weapons in the 1950s.

Legitimate concerns about the effects of radioactivity on the environment in general and on human health in particular were expressed by the public. The dangers of radiation from testing nuclear weapons became evident, and all countries except France stopped atmospheric testing by the late 1950s.

While one environmental problem had been "solved," by the early 1960s other environmental problems were becoming clear. The widespread use of pesticides and other contaminants took the place of radioactive fallout as an environmental issue. In the early 1960s, under the stimulus of President John F. Kennedy, many Americans were seeking ways to create a better world. The pollution of the environment gave many of them the cause they were looking for.

Within this framework for environmental change, Rachel Carson, a nature lover, wrote *Silent Spring,* a book that set the stage for the new crusade: the case against pesticide spraying. She wrote:

> The most alarming of all man's assaults upon the environment is the contamination of air, earth, rivers, and sea with dangerous and even lethal materials. This pollution is for the most part irrecoverable; the chain of evil it initiates not only in the world that must support life, but in living tissues is for the most part irreversible. In this now universal contamination of the environment, chemicals are the sinister and little recognized partners of radiation in changing the very nature of the world—the very nature of its life.

Of the use of pesticides, Carson wrote:

> Chemicals sprayed on croplands and forests or gardens lie long in soil, entering into living organisms, passing from one another in a chain of poisoning and death. Or they pass mysteriously by underground streams until they emerge and through the alchemy of air and sunlight, combine into new forms that kill vegetation, sicken cattle, and work unknown harm on those who drink from once pure wells. . . . The chemicals to which life is asked to make its adjustments are . . . the synthetic creations of man's inventive mind.
>
> These sprays, dusts and aerosols are now applied almost universally to farms, gardens, forests, and homes—nonselective chemicals that have the power to kill every insect, the "good" and the "bad," to still the songs of birds and the leaping of the fish in the streams, to coat the leaves with a deadly film, and to linger on in soil—all this though the intended target may be only a few weeds or insects. Can anyone believe it is possible to lay down such a barrage of poisons on the surface of the earth without making it unfit for all life?

Rachel Carson was a literary master who caught the imagination of the public. She had previously developed a reputation as a naturalist with such books as *The Sea around Us* and *The Edge of the Sea. Silent Spring* became a classic because it brought to the public's attention a subject that had received little attention previously—pesticides—and it attacked government and scientific agencies for not providing protection for this insidious danger. The environmental movement now had a focus—one that persists to the present day.

Although *Silent Spring* contained some scientific inaccuracies and some distortions of scientific fact, it served as a catalyst, sounding the alarm and reminding people that a clean environment is a precious commodity that must be protected from human carelessness and irresponsibility. The stage was now set for the first major federal environmental legislation on hazardous material and toxic waste. The public entered directly into the environmental discussions through such organizations as the Sierra Club, the Environmental Defense Fund, and the National Audubon Society.

Public awareness of the pollution of the environment began to increase in the 1960s. Because this was an evolutionary process, the following discussion presents some of the major events as they occurred. The fear of pesticides was a powerful stimulant of public reaction in the 1960s. In the early 1970s, the discovery of toxic dump sites across the nation crystallized thinking that environmental pollution had achieved national proportions. This led to the Resource Conservation and Recovery Act of 1976 (RCRA), which established the basic framework for conserving our resources. The last portion of this chapter discusses some of the toxic chemicals, techniques developed to remove them from the environment, and the gradual recognition that the problems are far greater than originally thought.

Generation of Wastes

Hazardous and toxic wastes are the "garbage" of a highly technological society. The United States generates approximately 4.5 billion tons of solid wastes annually, or roughly 100 pounds per person each day (Table 1.1). These wastes are produced by many segments of our society. Agriculture is the largest producer, with 2.3 billion tons, or 52 percent of the total. This is followed by the mineral industries, with 36 percent of the total; industrial waste (nonhazardous), with 5 percent; domestic, municipal waste, about 4 percent; and utilities, about 2 percent. Thus, about 99 percent of the waste generated is not hazardous or toxic: only about 45 million tons, or 1 percent, can be classified as dangerous to humans or the environment. Of the generators of hazardous and toxic wastes, industry produces nearly 100 per-

TABLE 1.1
Solid Waste Generated Annually in the United States

SOURCE	MILLIONS OF TONS	PERCENTAGE
Agriculture	2,340	52
Mineral industries		
(mining and milling waste)	1,620	36
Industrial		
(nonhazardous)	225	5
Municipal		
(domestic)	180	4
Utility	90	2
Hazardous	45	2
Low-level radioactive	3	0.0007
TOTAL	4,543	100

Source: U.S. Department of Commerce, 1988

cent of the total. Of the industrial producers of hazardous wastes, chemical and allied products generate 60 percent of the total, followed by machinery, primary metals, paper and allied products, fabricated metal products, and stone, clay, and glass products.

Definition

Hazardous waste may be defined as solid or liquid waste (or a combination of solid and liquid wastes) that, because of its quantity, concentration, or physical/chemical or infectious characteristics, may (1) cause or significantly contribute to an increase in mortality or an increase in serious irreversible or incapacitating reversible illness or (2) pose a substantial present or potential hazard to human health or the environment when improperly treated, stored, transported, disposed of, or otherwise managed.

The U.S. Environmental Protection Agency identifies as hazardous any waste that possesses one of the four characteristics listed below. These wastes are subject to regulation under the Resource Conservation and Recovery Act.

Ignitability. This characteristic identifies wastes that pose a fire hazard during routine management. Fires not only present immediate danger of heat and smoke, but can spread harmful particles over wide areas. A waste is ignitable if it is (a) a liquid solution which contains less than 24 percent alcohol by volume and has a flash point less than 60C, (b) not a liquid solution but under standard temperature and pressure may cause a fire through friction, or (c) an ignitable compressed gas containing an oxidizer, such as peroxide.

Corrosity. This term identifies wastes requiring special containers because of their ability to corrode standard materials, or requiring segregation from other wastes because of their ability to dissolve toxic contaminants A waste is corrosive if it is a liquid and has a pH less than or equal to 2 or if it is a liquid which corrodes steel at a specified rate.

Reactivity (or explosiveness). Reactivity identifies wastes that during routine management tend to react spontaneously, to react vigorously with water, to be unstable to shock or heat, to generate toxic gases such as cyanide or sulfide, and finally to be capable of deterioration.

Toxicity. Toxicity defines wastes that, when improperly managed, may release toxicants in sufficient quantities to pose a substantial hazard to human health or the environment.

The Environmental Protection Agency has prepared a list of 363 compounds that it classifies as hazardous. This list of organic and inorganic substances is subject to changes as new tests for hazardous materials are developed.

Evolution of Pesticide Controls

Pesticides, or agrochemicals, are chemicals designed to control various pests that interfere with the growing of agricultural and horticultural crops. They fall into three major categories: insecticides, herbicides (vegetation control), and fungicides. Since the dawn of agriculture, humans have continually endeavored to improve crop yield. This has led to attempts to control or eliminate insects and diseases that ravage crops. Sulfur was used to control insects and crop diseases before 100 B.C., and its use as a fumigant was mentioned by Homer. Pliny in 79 A.D. advocated the use of arsenic as an insecticide, and in the sixteenth century the Chinese were applying arsenic compounds as insecticides.

Tales of crop destruction are among the earliest writings. There are many references in the Bible to the insect ravages of nations. These plagues were thought to emanate from God, who sent them to punish the evil actions of humankind. It was not until the nineteenth century that systematic scientific methods began to be applied to the problem of controlling agricultural pests.

About 1850, two important natural insecticides were introduced: rotenone, from the root of the dervis plant, and pyrethrum, from the flower buds of a species of chrysanthemum. Both are still widely used. Soap was used to kill aphids, and sulfur was used as a fungicide on peach trees. In 1867, Paris green (copper arsenic) was introduced to control the Colorado beetle in Mississippi. By 1900 Paris green was so widely used that certain states passed laws to control its use.

In the latter part of the nineteenth century the success of the simple chemical pesticides stimulated the search for new compounds of copper, mercury, sulfur, and others. In addition, the manufacture of equipment for effectively applying these materials had begun. Many of the well-known poisons began to be used in controlling insects and other pests. Cyanide, generally in the form of hydrogen cyanide gas, was used in the 1880s as a fumigant to kill insects in California citrus groves. As insects developed resistant strains, other poisons were used. In 1897 formaldehyde was introduced as a fumigant and in 1913 organomercurials were first used as fungicides against cereal smut diseases. In 1912, calcium arsenate replaced Paris green and lead arsenate in boll weevil control.

It was eventually found that vegetables sprayed with arsenical insecticides contained poisonous residues. By the 1920s the public outcry against the use of these insecticides was so great that a search began for less dangerous alternatives. This led to the introduction of organic compounds, such as tar, petroleum oils, and dinitro-o-cresol. This last compound was patented in 1933 as a herbicide (Sinox) to control weeds in cereal fields. Unfortunately, it was a very poisonous substance.

The 1930s witnessed the beginning of the modern era of synthetic organic pesticides, of which there are now hundreds. This began a period of rapid growth of pesticide use in the industrialized countries of the world, a development that affected not only the high costs and shortage of labor, but also the demand for a quality food product. As the populations of the Third World countries increased, and greater food supplies were needed, the use of pesticides grew in those areas as well.

Pesticide Eradication Programs

By the 1900s it was recognized that humans had the potential to eradicate whole species of insects that had spread over vast areas. In the United States it was decided that spraying would be used to eradicate the fire ant and the gypsy moth. These massive endeavors not only received public attention, they provided environmentalists with their first opportunities to protest the use of pesticides over areas of thousands of square miles.

Fire Ants

The fire ant was brought from South America to the United States on ships in the 1920s. The first ants appeared in the port of Mobile, Alabama, and soon migrated to nine southern states. Like most ants, the fire ant delivers a painful sting, but no greater than that of bees or wasps. The fire ants were a nuisance primarily because they built high mounds of dirt that interfered with farming.

In 1957, the Mississippi Farm Bureau requested the Agricultural Research Service (ARS) of the U.S. Department of Agriculture (USDA) to investigate the problem. The ARS also had input from a number of southern commissions of agriculture. Because of the influence of these groups, a bill to amend the Agricultural Organic Act of 1946 was introduced in Congress. The Fire Ant Eradication Act was enacted, with a minimum of debate or opposition, in less than three months. This bill gave the USDA broad authority to control or eradicate insect pests, particularly the fire ant, without specific congressional authorization.

As soon as the bill was passed, controversy over the spraying of thousands of agricultural areas arose. On May 23, 1957, the day President Eisenhower signed the bill into law, the *New York Times* editorialized on "poison in the air" and supported expanded federal research on the relationship between pesticides and wildlife.

In order to win public support for carrying out the fire ant control program, the USDA undertook a remarkable publicity campaign. The propaganda extolled the role of the government in controlling the "vicious" fire ant. John Devlin, using the USDA material, wrote in the *New York Times* on December 23, 1957: "Farmers find the ant costly. Mowing bars hitting mounds jam and break. Because the ant tunnels several feet underground, poor soil is also distributed on top of the topsoil. Furthermore, the ants attack and kill calves, pigs, and chickens. They also bite farm workers." The publicity campaign was successful in the South.

While the campaign rallied southern leaders to join the control program, it brought attention to those critical of federal pesticide programs. Rachel Carson's chronicle of the fire ant program in *Silent Spring* infuriated those in charge of the program. To contradict much that had previously been written in the *New York Times*, the editorial board wrote on January 8, 1958: "There is good reason to get rid of the fire ant. . . . But the way in which the Agricultural Department is suddenly committing itself on such an enormous scale may be definitely worse than the fire ants' bite. It is rank folly for the government to embark on an insect-control program of this scope without knowing precisely what damage the pesticide itself will do to both human and animal life, especially over a long period." This criticism did not deter the program's supporters. It did, however, spur congressional reaction to pesticide research legislation that had languished since its introduction at the height of the gypsy moth episode. This program was only mildly successful, but the stage was set for the onslaught against the use of pesticides that erupted in the 1960s.

Gypsy Moth Campaign

Gypsy moths strip trees clean of leaves, and by the middle 1950s they were a widely acknowledged threat to U.S. northeastern forest regions. In 1957, at about the same time as the attack on the fire ants in the South, the U.S. Department of Agriculture decided to spray the forests of the Northeast. The Division of Plant and Tree Disease expressed confidence that the gypsy moth could be eradicated. It was recommended that the USDA, in conjunction with four northeastern states, aerially spray three million acres of infested forest with a mixture of DDT suspended in oil. Such a broad-based approach would be more effective and cheaper than the traditional method of spraying individual trees.

Elected officials in affected states gave enthusiastic support, and in April 1957 planes sprayed wide swatches of land. Little advance notice had been given to the inhabitants, for the USDA felt no harm could be done to humans by the chemicals. On May 8, however, a group of Long Island residents led by Robert Cushman Murphy, the noted ornithologist, petitioned the New York Federal District Court to stop the USDA spraying. The enjoinder claimed that the DDT mixture in use was far too heavy for the intended purpose, and that the planes used for spraying violated aviation guidelines by flying too low and during high winds. The group claimed that local wildlife and gardens would be harmed by the spraying. In addition, hundreds of farmers claimed that crops were damaged. On May 22, Governor W. Averell Harriman stated that crops were damaged and urged greater care in spraying. The U.S. Department of Agriculture acknowledged that some mistakes had been made.

At the same time, New York residents were reporting fish mortality throughout the state. Nevertheless, the USDA persisted in saying that no harm would come to the fish population as a result of the spraying. As a response, the *New York Times* questioned the government's moral and legal right to "spread poison" over private property without permission of the owner. They also cited the U.S. Fish and Wildlife Service's misgivings about the spraying campaign. In May 1957, the federal court denied the request of the Long Island residents for an injunction, indicating that insufficient evidence was presented that DDT posed a clear and immediate threat to wildlife or human health.

The gypsy moth spraying concluded in June 1957. While evidence remained inconclusive as to the threat posed by DDT, a regional awareness was generated that began the long campaign against its use. The program was a critical catalyst for the priorities of established conservation organizations, which had previously paid little attention to the potential harmful impact of chemicals on wildlife. Of these organizations, the National Audubon Society was perhaps most affected; its goals and priorities were redirected and revitalized. In the 1900s the Audubon Society directed its attention to preserving wetlands and forest areas as bird and wildlife sanctuaries. As reports indicated the decline in populations of falcons, bald eagles, and ospreys, the group's concerns were broadened. At first pesticides were not linked to these phenomena, but the gypsy moth campaign transformed the Audubon Society's ambivalence toward chemical sprays. In essence, the gypsy moth campaign changed the Audubon Society from a "bird club" to an "environmental protection organization."

Environmental Focus on DDT

In the 1960s the campaign to ban DDT became the environmentalists' starting point for the control of insecticide spraying. DDT (dichloro-diphenyl-trichloro-ethane) was the first pesticide to be challenged as an environmental and health hazard. It was first synthesized by a German chemist in 1874, but its properties as an insecticide were not discovered until 1939. Dr. Paul Müller, a Swiss chemist who was awarded the Nobel Prize in medicine in 1948, patented it as a control insecticide in 1939. DDT replaced a large number of poisonous chemicals, such as mercury, arsenic, lead, and fluorine, in the control of insects during the 1940s and 1950s.

DDT had an immediate effect on the health of the world. Prior to 1940, more than 200 million people worldwide were stricken annually with malaria, with an average of 2 million deaths a year. By the 1950s DDT had drastically reduced the cases and deaths from malaria. Initially DDT was used only to control malaria, but its success in this area led to its use to control yellow fever, sleeping sickness, plague, typhus, and encephalitis, all of which are transmitted by insects. Its low cost and effectiveness led to its later use in agriculture to control insects. As a result, crop yields increased in many areas of the world.

Of the insecticides available in the 1950s, DDT dominated the market. There were three reasons for its wide acceptance. First, it was the least expensive insecticide available. Second, it exhibited an exceptional stability and persistence after spraying, which reduced the number of treatments needed in a given area. This characteristic was particularly valuable in control of mosquitoes in curbing malaria. Third, DDT combined high toxicity for insects with low toxicity for other forms of life, including humans. The recognition of the great contribution of DDT to human welfare was worldwide. In 1969, the *British Medical Journal,* in an article titled "DDT in the Environment," stated that DDT was a "miracle" chemical that "has been incontrovertibly shown to prevent human illness on a scale hitherto achieved by no public health measure entailing the use of a chemical."

While DDT was being acclaimed as the wonder pesticide, entomologists, biologists, and toxicologists were beginning to question its safety. In the early 1960s a number of deaths were attributed to DDT. In 1963, however, the *Reader's Digest* reported, "Half of the deaths were not due to improper use at all, but were accidents in which children got hold of the toxic material. The other deaths have been traceable to careless handling and application of some pesticides."

In 1962, with the publication of *Silent Spring,* the debate reached fever pitch, sparked by the fact that DDT accumulates in the body. As Rachel Carson wrote:

> Once it has entered the body it is stored largely in organs rich in fatty substances (because DDT itself is fat-soluble) such as the adrenals, testes, or thyroid. Relatively large amounts are deposited in the liver, kidneys, and the fat of the large, protective mesenteries that enfold the intestines. . . . Because these small amounts of pesticides cumulatively are stored and only slowly excreted, the threat of chronic poisoning and degenerative changes of the liver and other organs is very real.

As a response, the environmental movement began a campaign to ban the use of DDT. The Environmental Defense Fund was organized. The attack on DDT was based on three main arguments. First, DDT was causing large decreases in and perhaps extinction of certain bird populations because it caused the birds' eggshells to weaken and the eggs were often destroyed before the birds could hatch. Second, DDT was so stable that it could not be removed from the environment in a reasonable length of time and was thus hazardous to the ecologic balance. Third, DDT was a possible cause of human diseases, including cancer.

In the 1960s the overwhelming scientific evidence was that DDT was safe and effective. The National Academy of Sciences reported:

It is estimated that in little more than two decades, DDT has prevented 500 million deaths that would otherwise have been inevitable. (*Life Science,* 1971)

Dr. Jesse L. Steinfeld, surgeon general, U.S. Public Health Service, stated:

DDT has been instrumental in literally changing the course of history for many nations and continues to do so today. Its use . . . has meant the difference between hunger, despair, and poverty and food, hope and the promise of a better life to billions of people throughout the world. (testimony before the Environmental Protection Agency, 1971)

In spite of the many favorable scientific reports, the crusade against DDT grew in the late 1960s. In 1969, Michigan was the first state to ban the use of DDT, followed by Wisconsin, Arizona, and California. That same year, the Environmental Defense Fund, the Sierra Club, the National Audubon Society, and a Michigan environmental group petitioned Secretary of Agriculture Clifford Hardin to ban DDT, alleging it was a carcinogen. Also in 1969, a commission on pesticides and their relationship to environmental health was appointed by the secretary of the U.S. Department of Health, Education and Welfare. Based on the commission's report that DDT is harmful, the secretary of agriculture developed a plan for phasing out the use of DDT by December 31, 1970, with the exception of uses essential to protect public health and welfare.

This partial ban did not satisfy environmentalists, and they petitioned the newly formed Environmental Protection Agency for a total DDT ban. In 1971 and 1972 hearings were held. In April 1972, Edmund Sweeny, chairman of the hearings, after receiving 9,300 pages of testimony and more than 300 technical documents, recommended to the EPA that "no more extensive ban of DDT was necessary or desireable; based on evidence of the hearings." DDT was recognized as a major contributor to human welfare.

Less than two months after the Sweeny report, William Ruckelshaus, administrator of the EPA, banned all of the remaining uses of DDT with the exception of uses related to essential public health purposes. The banning of the use of DDT by Ruckelshaus rejected all of the Sweeny report's recommendations. It appears that the EPA decision disregarded the overwhelming opinion of the scientific community.

The decision to ban DDT was contested in the federal courts, and in December 1973 the ban was upheld. By this time DDT was not the leading insecticide in the war against pests, but it remained the most important symbol of the pesticides paradigm. It reflected human ability to control the environment. The defenders of DDT pointed out its great advantages in improving crop production; many felt that the law would limit or even prevent the use of any pesticide. On the other hand, zealous environmentalists felt that the potential dangers to health and the environment were sufficient justification to ban all pesticides, regardless of the consequences of lowered food production. By the 1970s, an alarmed public, the passage of the National Environmental Protection Act, and the creation of the Environmental Protection Agency combined to shift the tide from those traditionally in control of pesticide policy-making to the environmentalists. There was now an alliance between the environmentalists and control by government agencies.

Herbicide Control of Vegetation

Dioxin

In the late 1960s the attack on the use of herbicides was added to that on insecticides by the environmentalists. Of these, the phenoxy acid compounds, some of which are contaminated with dioxin, a deadly substance, were of major importance. The use of this herbicide to control undesirable weeds had become of major importance in crop production.

There are about 75 types of phenoxy acid compounds, many of which do not contain dioxin. The most popular of these is 2,4-dichlorophenoxyacetic acid, or 2,4-D, because of its low cost and effectiveness. However, for those shrubs and plants that are resistant to 2,4-D, other herbicides such as 2,4,5-T are substituted. However, 2,4,5-T has dioxin as an unavoidable contaminant.

In the early 1970s the use of 2,4,5-T was linked to human miscarriages in the Alsea area of Oregon. In addition, two epidemiological studies were conducted in Corvallis, Oregon, in a sparsely populated rural area near the Idaho border. The 1979 report concluded that "the Agency's systematic survey of the occurrence of spontaneous abortions in areas of 2,4,5-T use indicates that there was an unusually high number of spontaneous abortions in the area, and that the incidence of spontaneous abortions may be related to 2,4,5-T in that area." Although the findings of the report were questioned by a number of scientists, the Environmental Protection Agency in 1979 issued an emergency suspension of all remaining uses of 2,4,5-T except for range-land clearing and weed control in rice fields.

Agent Orange

The spraying of 2,4-D and 2,4,5-T beginning in 1962 to destroy vast areas of vegetation in Vietnam drew the attention of the general public to the use of herbicides to a greater degree than any other event. There were reports of chemical killing of vegetation in essentially every newspaper in the United States. Many of the statements were highly emotional. For example:

> An ultratoxic herbicide was rained down leaving an endless harvest of genetic defects and cancer; tens of millions of Vietnamese and 2.4 million American soldiers are estimated as having been contaminated with Agent Orange in Vietnam. (Karl Grossman, *The Poison Conspiracy*. Sag Harbor, NY: Permanent Press, 1983, p. 55)

> The enemy is 2,4,5-T, a powerful phenoxy herbicide contaminated with dioxin, generally considered the deadliest substance ever created by chemists. The story of 2,4,5-T symbolizes many environmental issues, pitting a powerful corporation against a grassroots movement unaided by the government agencies designed to protect the environment and the public health. (Ralph Nader et al., *Who's Poisoning America*. San Francisco: Sierra Book Club, 1981, p. 240)

Due to the extremely adverse publicity on the use of herbicide spraying, the U.S. Department of Defense stopped the use of all herbicides in 1970 for the defoliation of the jungles in South Vietnam. At that time the Defense Department considered the use of Agent Orange a closed issue.

The belief that cancer could develop in soldiers exposed to Agent Orange (the herbicides were stored in color-coded drums from which the name was derived) developed from a phone call in 1976 made by the wife of a dying Vietnam veteran to the Veterans Administration's Chicago office. A VA counselor, Maude deVictor, began to keep records of ailments reported by Vietnam veterans, ranging from liver disorders to miscarriages suffered by veterans' wives. She reported her results to CBS TV in Chicago, and in March 1978 the documentary *Agent Orange: Vietnam's Deadly Fog* was broadcast. In July, ABC reported a similar story on the television program *20/20*. Thus began the Vietnam saga of the health effects of Agent Orange.

In September 1978 Paul Reutershan, a Vietnam veteran dying of liver cancer, filed a $10 million federal lawsuit claiming Agent Orange caused his cancer. Later in 1978, Victor Yannacone, Jr., filed a class action suit indicating that 40,000 veterans and their families may have been contaminated by Agent Orange. The suit was filed against five manufacturers of chemical compounds—Dow, Monsanto, Hercules, Diamond Shamrock, and Thompson-Hayward.

On March 31, 1983, Bruce E. Herbert, deputy director of the Center for International Security, wrote in the *Chicago Tribune*:

> After nearly five years of almost constant publicity, as of March 1, 1983, only 16,821 veterans have ever filed claims with the VA for suspected Agent Orange damage. Of this number, less than 8,400 present any certifiable medical condition, whether or not these disabilities can be scientifically linked to exposure of Agent Orange. Three thousandths of one percent of the 2.4 million men who could have been exposed to Agent Orange in Vietnam is hardly a compelling statistic on which to make assumptions about "unusually large numbers" of veterans suffering latent Agent Orange-induced health impairments.

In spite of this evidence the court trial against the chemical companies was set for May 7, 1984. Judge Jack B. Weinstein warned both sides of the problems they faced in the impending trial. The judge indicated to the companies and insurance agents that because of the emotions attached to the trial, in spite of scientific evidence that no medical problem existed, they were likely to lose the case. At the same time, the Brooklyn jury was sympathetic to the veterans, stacking the odds even higher against the chemical companies. Before the trial started, an out-of-court settlement was reached totaling $180 million.

The reports on dioxin as a cause of cancer are conflicting. In 1977 a Swedish study found a relationship between the use of phenoxy herbicides and chlorophenols and the development of a soft-tissue sarcoma (STS). In another study in 1983 by the U.S. National Institute of Occupational Safety and Health (NIOSH) of more than 4,000 workers who manufactured phenoxy herbicides and chlorophenols, it was found that the sample population that had been exposed to the dioxin-containing products for 10 to 20 years had shown some evidence of chloracne. NIOSH reported 5 cases of STS, which statistically amounts to 21 times the expected number of cases.

In contrast, a 1982 New Zealand study that focused on 102 STS patients found that between 1976 and 1980 there was no excess for the occupational group involving agriculture and forestry, in spite of the fact that phenoxy herbicides had been used extensively for many years in New Zealand (Smith et al., "Do Agricultural Chemicals Cause Soft Tissue Sarcoma? Initial Finding of a Case-Control Study in New Zealand," *Community Health Studies* 6 [1982]: 114).

Possibly the most significant studies to test the health effects of this herbicide as a carcinogen were the so-called Ranch Hand investigations. These were named after the Air Force spraying teams that sprayed Agent Orange in Vietnam. Some 1,200 men were engaged in these flights. Many of the flyers reported that Agent Orange accumulated on their clothing during the spraying. The two studies that were carried out on veterans could not find statistically significant differences in health status between those who worked daily with Agent Orange in Vietnam and those veterans who did not. These studies refute the claim that Agent Orange was to blame for a host of cancers, birth defects, and other chronic illnesses on exposure to the defoliant.

The public has been deluged with stories about the effects of dioxin exposure. However, there are still major questions about the relationship between dioxin and human health. There can be no doubt that dioxin can kill, but many distortions and misconceptions are still present. On May 31, 1983, the *Wall Street Journal* brought the importance of dioxin as a health hazard into proper perspective in a column titled "Dioxin Hysteria":

> Clearly, there are reasons to be worried about the stuff. High concentrations in the soil—30,000 times greater than the CDC (Centers for Disease Control) evaluation standard—killed horses at Times Beach. . . . What is known, however, suggests that most of the scare stories are exaggerated. Human exposure to dioxin has so far been scientifically linked to only one health problem—a skin disease called chloracne, which tends to go away fairly rapidly. . . . The notion that dioxin is a doomsday menace is based less on medical evidence than on some kind of psychological phenomenon.

Pesticide Regulation

The original pesticide law, Federal Insecticide, Fungicide and Rodenticide Act of 1947 (FIFRA), required that pesticides be registered by the secretary of agriculture before being marketed, in order to assure that they were effective for the purpose stated. Registration was thus basically a license detailing the specific use of the chemical. The law was primarily concerned with the performance of the chemical, and not with the protection of public health. It was not until 1967 that the Pesticides Registration Division in the Department of Agriculture initiated its jurisdiction to recall a dangerous pesticide.

In 1970 the administration of the federal pesticide law was transferred from the Department of Agriculture to the newly created Environmental Protection Agency. In 1972 Congress passed the Federal Pesticide Act, with amendments in 1975, 1978, and 1980. The new law and later amendments refocused emphasis on the safety of pesticides and on making the registration process for chemical pesticides more rigorous through tests that would reveal their potential to cause cancer, birth defects, and genetic mutations.

The Environmental Protection Agency has the authority to approve all new pesticides before they can be sold or used, as well as new uses for old pesticides. In addition, EPA is required, under the Federal Food, Drug and Cosmetic Act, to establish tolerances for pesticides that remain in or on foods or animal feeds. The EPA was also given the responsibility of reexamining and reregistering all pesticides that were in existence when the FIFRA was enacted in 1947. This was necessary because the thousands of pesticides that already existed when the EPA was established had not been tested for their health and environmental effects according to the new laws. In essence, the Environmental Protection Agency is required to evaluate every pesticide by balancing the benefits of increased food production against the potential harm it might cause in humans and/or the environment. If a pesticide is judged to cause unreasonable risk or harm to humans or the environment, then its use is to be either restricted or banned.

When the Environmental Protection Agency was created it had the formidable task of reviewing 35,000 pesticides in use and establishing new tolerances for those that left residues in food. The task was to be completed by 1976, but it proved so formidable that in 1978 Congress eliminated the deadline and requested simply that it be done "as expeditiously as possible."

In order to expedite the process, the 1978 amendments to the pesticide law permitted a chemical-by-chemical rather than a product-by-product approach to the registration process. As a result, the EPA could assess the approximately 600 basic chemical pesticide ingredients common to the known 50,000 pesticide products. Nevertheless, a study by the U.S. General Accounting Office in 1986 found that the EPA had failed to evaluate the majority of the chemical pesticides in use. At the current pace the reassessment of the safety of pesticides will be completed sometime after the year 2000.

By March 31, 1986, preliminary assessments had been completed on 124 known active ingredients. Not a single final assessment had been completed on the 600 active pesticide ingredients. The preliminary assessments included cataloging data on the chemicals, identifying data gaps, and requesting missing or inadequate health and environmental studies from pesticide firms. Based on the initial data, some restrictions were placed on the use of 75 of the 124 chemicals evaluated, but none was completely banned from use. The reviews of each ingredient have normally taken from two to six years. During review, the chemicals continue to be used.

The review process has been complicated by the finding that "inert" ingredients in pesticides are sometimes toxic. Initially, the EPA evaluated only "active" ingredients as to their toxicity, but the agency has now identified more than 100 inert ingredients with known or suspected toxicity and an additional 800 for which data are insufficient to determine toxicity.

The evaluation of pesticides has experienced two major problems. First, there has been a failure in enforcement of the legislation. The review procedure has been remarkably slow and minimal. A basic reason for this is a lack of resources. Between 1980 and 1986 the personnel in the pesticide program were reduced from 829 full-time positions to 591. The reduction in staff was, however, partly due to EPA policy that did not give pesticide control a high priority. As EPA Administrator Lee Thomas stated in 1985, we are "confident we have enough to do the job assigned to us." He continued, we "seriously doubt we could efficiently use any more." During the Reagan administration the EPA was committed in general to a hands-off political ideology concerning regulation of the chemical industry.

The second factor leading to slow progress in registering pesticides lies in the tradition of the 1950s, when the viewpoint prevailed that pesticides were essential to agricultural productivity. Neither the pesticide legislation nor the resulting regulatory program to enforce the law provides alternatives to the use of chemicals in agriculture. Integrated pest control, biological controls, and inorganic farming are considered minor contributions to the mainstream of chemical agriculture. There is thus a bias for the use of pesticides built into the enforcement program. There is no inherent intent in the law to minimize the use of chemical pesticides and to maximize the use of ecological agriculture. Rather, the law provides a safe venue for chemical pesticides to develop and be used in the market in a liability-conscious society.

Pesticides in the World

The use of pesticides throughout the world depends on the level of economic development, political climate, and ecological setting of a nation. A major division appears between the developed and developing countries. In the advanced nations, laws and regulations based on health and environmental criteria are increasing. In contrast, in developing nations, where food supplies are precarious, regulations are based on agricultural development criteria.

The effective use of agrochemicals in developing countries depends to a large extent on the availability of government expertise and advice to farmers through extension services, rather than new pesticides. Misuse of pesticides can be counterproductive as well as hazardous. Thus, in developing countries use of pesticides requires administrative resources and an industrial program. In the developed countries the security of supply of both agrochemicals and agricultural products has advanced from policy debates toward consideration of the costs of pesticide use.

A developing global consideration has also become increasingly pronounced in recent years. New questions are being asked, such as, How much pesticide has accumulated in the biosphere? What risks are involved in this process? Questions are also being raised concerning pesticide residues in agricultural imports from Third World countries. Clashes based on competing values have also arisen. For example, attempts are now being made to define more precisely the responsibilities of parties using pesticides or other toxic chemicals, especially if that use results in damage to others.

In July 1990 a survey by the American Farm Bureau revealed that more than half of all Americans were "very concerned" about farm pesticides on the food they eat, and one in five would like to see agricultural chemicals banned. Further, the survey found that three-fourths of the public believes farmers use more agricultural chemicals than they should. Almost half of the public is very concerned about growth hormones given to livestock to fatten them. There is, then, strong evidence that Americans continue to be concerned about the safety of their food supply. Dean Kleckman, Farm Bureau president, called the new poll results "disconcerting" for U.S. farmers, who he believes produce the safest food in the world. Kleckman does not believe that the answer to this problem is more stringent government regulation and limits on chemical use, as proposed by some lawmakers for the 1990 farm bill. Instead, the Farm Bureau plans to mount a public relations campaign to educate consumers about the need for farm chemicals and the safety of the U.S. food supply. The struggle to determine the volume of pesticides allowable by weighing the positive results against the negative is still in its infancy.

Hazardous and Toxic Waste Sites

For many years the effects of toxic and hazardous chemical and biological by-products of our technological society were ignored. Abandoned dump sites are the most prominent feature of these practices. Growing awareness of these problem sites has led many Americans to conclude that hazardous waste management is among the most serious of the environmental problems facing the nation. However, the problem is so recent that state and national policies are only now being established.

Major Hazardous Waste Sites

In the 1970s a number of the areas where major deposits of hazardous and toxic materials had been stored or disposed of began to create environmental and health problems. Because the sites were located across the United States, national attention was focused on them. The following sections describe selected examples of these major hazardous waste sites.

Love Canal, Niagara Falls, New York

The saga of the Love Canal began in 1892, when William T. Love proposed to build a power canal along the Niagara River starting about eight miles above Niagara Falls. The canal would carry water to the edge of the cliff, where it would drop about 1,300 feet, producing electricity. The canal was partially dug when the project was abandoned. For many years thereafter, children swam in the canal in summer and skated on its surface in winter.

In the 1940s, the Hooker Chemical Company (which in 1968 became affiliated with the Occidental Petroleum Company) bought the canal and, until 1953, used it to dispose of waste materials. The canal was considered an excellent disposal site; it was about 10 feet deep and 60 feet wide and was dug into a clay layer, a material through which liquid flowed slowly. It is estimated that 20,000 metric tons of waste in metal drums were buried and covered with clay. The clay was intended to create a "vault" to hold the chemicals securely.

By 1953 a large portion of the canal was filled when the Hooker Chemical Company sold the land to the Niagara Falls Board of Education for $1.00. In the deed it was stated that the property was filled with waste products from the manufacture of chemicals, but no warning was given that a potential health problem existed. A school was built on the rim of the canal, and the remainder of the land was sold for construction of homes.

The first evidence of health problems emerged in 1958, when children received chemical burns while playing around the road-building operations. It was discovered that a few of the drums holding the waste had ruptured. The waste material was removed and the area refilled, apparently solving the problem.

The modern-day Love Canal problem can be traced to 1976, when residents whose homes were adjacent to the canal began to complain of chemical odors from the landfill and claimed that their discomfort was related to the seepage of wastes from the Love Canal property. To emphasize the health problem, reporter Michael Brown of the *Niagara Gazette* began a series of articles about suspected cases of toxic waste-induced illnesses among people living in the Love Canal area. The *Courier Express–Niagara News* also contributed to an atmosphere of concern.

In 1978 Love Canal began to receive national publicity when the New York State commissioner of health began to investigate the situation. Eventually, more than 400 chemicals were identified in the area, including a number of carcinogens, teratogens, and mutagens. Some were measured at 5,000 times the maximum safe concentrations. Health problems began to be reported. Women located in certain sectors of the canal area suffered miscarriages at the rate of 50 percent higher than normal. In one sector, 4 of 24 children had birth defects. Many adults began to suffer liver damage. In August 1978 pregnant women and children under 2 years of age were encouraged to leave the area.

President Jimmy Carter declared Love Canal a disaster area, which made federal aid available to the residents. Initially, 237 homes were purchased by the state and boarded up. The school was closed and a chain-link fence was put up around the area. In May 1980 President Carter declared a second emergency, and an additional 710 families were offered the opportunity to sell their homes to the government.

Tens of millions of dollars have been spent on the cleanup and purchase of homes in Love Canal. Lawsuits totaling billions of dollars have been filed by Love Canal residents against Hooker, the Niagara Falls Board of Education, and the city of Niagara Falls. Both New York State and the U.S. Environmental Protection Agency have sued Hooker. The lawsuits will not be settled for many years. Love Canal is a ghost town.

The events at Love Canal alerted the United States to the problem of toxic chemical dumps as no other incident had done before. The pioneering efforts to defuse Love Canal's "ticking time bomb" initiated investigations of other wastes that had been buried and almost forgotten, many of them potential health hazards.

In May 1990 the Environmental Protection Agency indicated that the danger from toxic chemicals was no longer present in most of the Love Canal area. The houses that had stood vacant for years would be sold at reduced costs to private citizens. However, this does not conclude the saga of the Love Canal; environmental groups, believing that a toxic danger still exists in the area, plan to protest the decision of the EPA in the courts.

Times Beach, Missouri

In the early 1970s, some 2,000 gallons of a combination of oil and industrial waste were sprayed on a horse farm near Times Beach, Missouri. The waste was contaminated with dioxin. Soon after the spraying, birds, rodents, and horses died in the area, and two children playing in the area required hospitalization. By 1975 it was apparent that the area was contaminated. To complicate the problem, the citizens of Times Beach were concerned that flooding from frequent thunderstorms would spread the toxic substances throughout the town.

The Centers for Disease Control (CDC) recommended limit of lifetime human exposure to dioxin, 1 part per billion (ppb), was exceeded by some of the soil samples, which increased to more than 100 ppb in some areas. The CDC officials recommended evacuation. The Environmental Protection Agency agreed to spend some $30 million to purchase all of the homes in Times Beach.

To date there is no evidence that the health of the people of Times Beach has been harmed. After the removal of the town's population, the Missouri director of the Division of Health stated, "We have seen nothing to alarm us or to make us believe that Missourians are feeling acute health effects." In June 1983, the American Medical Association passed a resolution at its Chicago convention that there is no scientific evidence that dioxin poses a direct threat to human health. It voted to "adopt an active public information campaign . . . to prevent irrational reaction and unjustified public fright and to prevent the dissemination of possibly erroneous information about the health hazards of dioxin" (*Washington Post,* June 23, 1983).

The Times Beach incident illustrates the hysteria a community can suffer when it becomes known that a toxic substance is located in a populated area. Times Beach also illustrates that reliable information is not always available to the public. However, when it is believed that toxic substances contaminate an area, steps must be taken quickly so that human health will not be adversely affected.

In July 1990 the Environmental Protection Agency announced a plan to clean up the abandoned Times Beach area. An incinerator will be built, at a cost of $118 million, to burn the waste material. When the cleanup is completed the incinerator will be dismantled. After more than a decade, Times Beach will once again become a habitable community.

Other Toxic Sites

There are numerous further examples of toxic sites. In Bayou Sorrel, Louisiana, millions of gallons of toxic waste were dumped in huge open pits. The air currents carried the pollution into nearby areas. In Saltville, Virginia, a chemical plant used open pits for more than three-quarters of a century to deposit mercury toxicants that contaminated not only the soil but the groundwater that flowed into the North Fork Holston River. In Byron, Illinois, more than 1,500 containers of such industrial wastes as cyanide, chromium, arsenic, and heavy metals were buried on farmland. As the containers leaked, the substances entered the soil and groundwater. At Shepherdsville, Kentucky, an area that came to be known as the "Valley of the Drums" had about 100,000 drums of waste deposited illegally on farmland. The drums, which had been abandoned there by chemical companies, gradually contaminated the soil and water supplies.

Although it can only be estimated how much waste is deposited in rural areas and small communities each year, unsafe pits, ponds, lagoons, and landfills litter the U.S. countryside.

Cleanup of Hazardous Waste Sites

By 1980 it was recognized that a national program was needed to focus on the cleanup of hazardous waste sites. The passage of the Comprehensive Environmental Response, Compensation and Liability Act (CERCLA), known as the Superfund, began the national effort.

Federal Superfund Program

The federal program to clean up hazardous waste sites began with high expectations of great success. This appeared reasonable; the plan had strong public and congressional support and a growing network of state, local, and private authorities. Further, the Environmental Protection Agency had led the nation in progress for clean air and water. With an appropriation of $1.6 billion, the task appeared relatively simple: to clean up what were assumed to be a few hundred discrete land-based toxic sites.

Regrettably, the program has been far more complex than originally visualized. To begin with, there was no proven technology on how to manage waste sites. The EPA had to devise ad hoc control strategies to deal with hundreds of unique problems. Progress has been slow. "As a consequence, the nation is now beginning to confront the real dilemma: how to reduce environmental risks from a growing list of sites presenting ever new complexities, in a situation characterized by incomplete knowledge, immature technology, and relentless pressure on a limited pool of resources" (Environmental Protection Agency, *Management Review of the Superfund Program*. Washington, DC: EPA, 1989, p. 2).

During the 1980s a contentious debate raged over Superfund policy and regulation by the EPA. Though the program has picked up speed, Superfund results have normally been below public expectations. Many citizens who expect a clean environment have increasingly questioned the commitment of the government. At the same time, Congress has established tighter and more ambitious targets for the Superfund.

Scope of the Problem

The actual number of hazardous waste sites, and their distribution, dwarfs the problem originally visualized. More than 31,000 hazardous waste sites have been identified, with 27,000 having undergone preliminary field review and classification. Of these, nearly 1,200 have been assigned high priority for further action by the EPA. An average of 2,000 sites are added annually. Still more are being cleaned up by states or private companies. Because remediation of the most damaged sites involves years of painstaking groundwater treatment, new sites continue to be placed on the National Priorities List (NPL) faster than existing sites are removed.

It is now recognized that the Superfund will continue for many years, and it will be costly. In 1989 the EPA estimated that the cost of remediation at the NPL sites is likely to be $30 billion, assuming half the work is to be done by the Superfund and half by responsible parties. It will take at least 13 years to remove the hazardous sites on the present list, and the agency expects new sites to be added at the rate of about 75 to 100 each year. According to current estimates, the National Priorities List is expected to grow to 2,100 sites by the year 2000. Currently, the average cost of cleanup per site is approaching $25 million, but this is expected to increase as the EPA's technology evolves.

The problem is large, complex, and long-term. It will require technical competence and ingenuity, creativity, political wisdom, and plain, old-fashioned persistence. Although the Superfund has grown to $8.5 billion, this amount is only a portion of what will eventually be needed to achieve final cleanup at all currently listed sites.

National Priorities List

Because the number of hazardous waste sites is far too large for them all to be removed at a single time, the EPA has established the National Priorities List, which designates the worst of the nation's known sites contaminated with hazardous substances. The government's guide to the establishment of the NPL is the National Contingency Plan (NCP), which outlines what authorities are responsible for abandoned or uncontrolled waste sites and defines criteria and methods for removal of wastes or for remedial purposes.

The implementation of the NCP involves an evaluation process with a report to a local, state, or federal authority of possible contamination. The assessment includes size of the site, contaminating possibilities, types and quantities of waste, local hydrological and meteorological conditions, and the environmental impact. At the site, evidence is gathered as to contamination, including effects on vegetation and quality of the water supply.

After evaluation, sites are ranked according to type, quantity and toxicity of waste, the number of people potentially exposed, the possible pathways for exposure, the vulnerability of local aquifers, and possible other factors. The sites are ranked by the Hazard Ranking System (HRS) to determine their eligibility for placement on the National Priorities List. This ranking system does not determine if cleanup is possible or the amount of cleanup needed. Rather, it measures the severity of the problem and the likelihood and potential magnitude of exposure to hazardous substances for humans and the contamination of the environment. This scoring system allows priorities to be made among the many thousands of hazardous waste sites.

The initial assessment and site evaluation provide the basic data for the system. From these data three scores are obtained that measure the possibility of (1) hazardous substance spreading from the site through groundwater, surface water, or air and reaching populated regions, (2) people coming in direct contact with hazardous substances, or (3) fire and explosion caused by hazardous substances. The first score is used to place sites on the National Priorities List and is generally called the HRS score. Sites are placed on the NPL only if they receive a ranking score of 28.5 or more (on a scale of 0 to 100), except when designated by a state as a priority site regardless of the score. The second and third scores are used to identify sites that need removal actions.

As hazardous sites have been identified, the NPL has grown steadily. In October 1981, the EPA published an interim priority list of 115 sites; it followed with an expanded list of 45 additional sites in July

1982. In December 1982, the agency published a list of 418 sites. This number had grown to 747 by 1989 (Table 1.2). The list is not complete, and many more sites will be added in future years.

Progress to Date

By 1989 the Superfund had removed 26 hazardous waste sites from the National Priorities List, with another 10 sites nearly completed. This does not appear to be an impressive record, but evaluating the work of the Superfund by tallying completed sites is inherently misleading. Years of work must first be done, in multiple stages, before the program is completed. The EPA monitors sites long after cleanup standards have been achieved to ensure that the standards remain in effect.

Success of the Superfund is more appropriately measured in terms of the successive interim steps that quickly provide a margin of safety for local residents. For example, the EPA has conducted emergency removals to attack the most immediate sources of toxic exposure at more than 1,300 sites in communities across the nation. This program cost more than $400 million, and the EPA has used its enforcement authority to get responsible parties to take removal action in another 200 sites. In addition, long-term cleanup work is currently under way to neutralize the sources, contain the spread, and systematically reduce or eliminate toxic pollution at more than 400 sites under Superfund's remedial program. In 1989 the EPA undertook major technical planning work at an additional 275 sites.

Strategies for the Superfund

In order for the Superfund to be ultimately successful, the Environmental Protection Agency must clarify, communicate, and gain broad acceptance for its fundamental strategies. Such an approach makes it possible to solve the riskiest problems first and then turn to remaining sources of long-term risk on a priority basis.

In 1989 the following proposed strategies were envisioned by an EPA task force:

1. Make Sites Safer—Control Acute Sites Immediately
   The EPA's first priority under Superfund is to stabilize hazardous waste sites and eliminate any immediate danger to citizens. This assessment will evaluate the environmental and public health impacts of each site and will be used to determine immediate action.

TABLE 1.2
National Priorities List Sites

	SELECTED FOR LIST	PROPOSED FOR LIST	TOTAL		SELECTED FOR LIST	PROPOSED FOR LIST	TOTAL
Region I				**Region VI**			
Connecticut	8	6	14	Arkansas	10	0	10
Maine	6	2	8	Louisiana	9	2	11
Massachusetts	22	1	23	New Mexico	6	4	10
New Hampshire	15	0	15	Oklahoma	8	3	11
Rhode Island	8	3	11	Texas	24	4	28
Vermont	4	4	8	**Region VII**			
Region II				Iowa	9	15	24
New Jersey	100	7	107	Kansas	9	2	11
New York	73	3	76	Missouri	14	7	21
Region III				Nebraska	3	2	5
Delaware	12	8	20	**Region VIII**			
Maryland	7	3	10	Colorado	13	3	16
Pennsylvania	71	24	95	Montana	8	2	10
Virginia	12	9	21	North Dakota	2	0	2
West Virginia	5	1	6	South Dakota	1	0	1
Region IV				Utah	5	7	12
Alabama	10	2	12	Wyoming	1	2	3
Florida	32	15	47	**Region IX**			
Georgia	7	6	13	Arizona	5	4	9
Kentucky	12	5	17	California	52	36	88
Mississippi	2	1	3	Hawaii	0	6	6
North Carolina	15	7	22	Nevada	0	0	0
South Carolina	14	7	21	**Region X**			
Tennessee	10	3	13	Alaska	1	0	1
Region V				Idaho	4	0	4
Illinois	23	16	39	Oregon	6	1	7
Indiana	30	7	37	Washington	25	17	42
Michigan	65	15	80				
Minnesota	40	0	40				
Ohio	29	3	32				
Wisconsin	35	4	39				

Source: U.S. Environmental Protection Agency, *An Analysis of State Superfund Programs: 50-State Study*, (Washington, DC: Office of Emergency and Remedial Response, 1989), pp. 49–52.

2. Make Sites Cleaner—Worst Sites, Worst Problems First
   The EPA will conduct the earliest remedial work for the
   most pressing problems.

3. Carefully Monitor and Maintain Sites over the Long Term
   The agency will monitor a site until it is certain that the site
   is safe. If the EPA selects a remedy where hazardous substances remain on-site, the EPA will conduct a review at
   least every four years after the initiation of remedial action.
   Based on the review, the EPA will protect human health and
   the environment.

4. Emphasize Enforcement To Induce Privacy-Party
   Cleanup
   The EPA will emphasize enforcement to induce potentially
   responsible parties (PRPs) to carry out more cleanups under
   EPA direction. The objective is to achieve enforceable
   agreements by PRPs to carry out cleanups or, when
   agreements cannot be reached, to order PRPs to carry
   out cleanups.

5. Develop and Use New Technologies for More Effective
   Cleanups
   In order to assure permanent remedies, the EPA is to seek
   innovative treatment technologies that reduce the toxicity,
   mobility, and volume of wastes at Superfund sites.

6. Improve Efficiency of Program Operation
   While the program is inherently long-term, some costly and
   time-consuming procedures at the sites can be reduced without loss of program quality. Site managers will seek greater
   procedural efficiency by using an "Our Program" concept.
   This means the EPA will match the most appropriate tools
   available to the environmental problems to be corrected.

7. Encourage Full Participation by Communities
   The EPA needs to reenergize its communication with
   communities in order to encourage broad-scale public participation in a constructive manner. The agency needs people
   directly affected by the program to share in realistic expectations and to assist in reaching proper decisions.

8. Foster Cooperation with State Agencies and with the
   Natural Resource Trustees and the Agency for
   Toxic Substance and Disease Registry (ATSDR)
   The EPA recognizes that other agencies have a significant
   role to play in the Superfund process, primarily the states
   and the Natural Resource Damage Trustees. Cooperation
   with the states is fundamental to Superfund success.

State Superfund Programs

Since the passage of the Comprehensive Environmental Response, Compensation and Liability Act of 1980, a partnership has developed between federal and state governments in the cleanup of hazardous waste sites. Coordinated cleanup efforts between federal and state authorities are currently treating numerous sites targeted by the National Priorities List. However, a vast number of known and important sites are not eligible for inclusion on the NPL, and states are now responsible for enforcing or funding cleanups at non-NPL sites. The prospect for increasing state involvement at both NPL and non-NPL sites depends on the willingness and capacity of states to develop effective programs, to secure adequate resources to fund cleanups, to obtain private funds for cleanups, and to conduct oversight activities. Under the Superfund Amendments and Reauthorization Act (SARA) of 1986, Congress required the EPA to involve states in the Superfund program in a "substantial and meaningful" way. The State and Local Coordination Branch (SLCB) of the EPA is responsible for developing regulations, guidance, and policies related to this congressional mandate.

At the present time 39 states have funded enforcement authorities, of which 25 are actively involved in managing removals and remedial actions at non-NPL sites. Some of the states also manage or oversee cleanups of NPL sites as well. Of the 39 states with funded authorities, 14 have funds replenished at specific times, resulting in high and low periods of activity.

The remaining 11 states have limited cleanup capabilities or enforcement authorities. Of these, 6 states have removal or emergency response programs but limited remedial action authorities or capabilities. Waste removal action may be needed in a variety of situations, including accidents during transport and at active hazardous waste facilities. In contrast, remedial action is intended to effect permanent or long-term solutions that consider two vital aspects of site remediation—source control and waste migration management—and meet legally applicable and appropriate requirements.

The remaining states do not have state Superfund programs per se, yet each addresses the problem of hazardous waste sites in some fashion. Nebraska uses its groundwater regulations, Oklahoma and Georgia use RCRA-type laws to order cleanups of hazardous sites, and Colorado has a fund but limits its use to CERCLA cost-share and related administrative costs at NPL sites.

Hazardous Waste Sites

The exact number of waste sites in the United States is unknown, and estimates within the 50 states vary greatly. The number of sites reported by a state is largely a function of the level of the state's program. Table 1.3 identifies waste sites as reported by the states; the numbers range from none in Nevada to more than 25,000 in California. Sites needing attention in states range from zero to 6,654 in California: 11 states have 100 or fewer sites needing attention, 16 have 100 to 300, 16 have 300 to 1,000, and 7 states have more than 1,000.

At least 20 states have compiled priority listings of hazardous waste sites, but they have not used any uniform system for establishing their priorities. Of the states that list priorities, 11 follow a formal ranking system using the federal Hazard Ranking System or another scoring system; 10 use a modified HRS or a nonquantitative ranking system.

To illustrate, South Carolina's State Priority List (SPL) contains all sites that do not qualify for inclusion on the National Priorities List. Maryland has compiled a Disposal Sites Registry, which is a listing of all ranked sites, including the NPL sites. Vermont combines all hazardous waste sites into one program.

A total of 28 states have developed inventories or registries of sites. These are broader than priority lists, and usually include unconfirmed sites. Connecticut's statute, for example, requires that a site be listed on the state inventory before funds are provided for cleanup. Ohio's informal list contains sites categorized after a preliminary assessment as high, medium, or low priority. Iowa uses a five-tier system ranging from "imminent threat" to "closed with no management needed." State budgeting has affected the development of site categorization.

State Statutes and Programs

Since the passage of the Comprehensive Environmental Response, Compensation and Liability Act, which authorized the Environmental Protection Agency to establish the Superfund program, many states have enacted laws and developed programs with authorities and capabilities similar to the federal Superfund program. These state programs have some or all of the following characteristics:

1. Procedures for emergency response actions and longer-term remediation of environmental and health risks at hazardous waste sites, including NPL and non-NPL sites
2. Provisions for a fund or other financing mechanism to pay for studies and remediation activities
3. Enforcement authorities to compel responsible parties (RPs) to conduct or pay for studies and/or remediation
4. Staff to manage publicly funded cleanups and oversee RP-local cleanups

TABLE 1.3
State Hazardous Waste Sites

EPA	TOTAL IDENTIFIED	SITES NEEDING ATTENTION	PRIORITY LIST	INVENTORY OR REGISTRY
Region I				
Connecticut	560	560		567[1]
Maine	237	117		317[1,2]
Massachusetts	1,800	1,725	1,152[3]	1,634[1]
New Hampshire	400	400		150–175[2]
Rhode Island	280	280		
Vermont	260	241	130[4]	50[2]
Region II				
New Jersey	3,225	3,000		336
New York	1,167	1,039	1,091	615[2]
Region III				
Delaware	200	160	48	200[3]
Maryland	304	254	251[1]	300[2]
Pennsylvania	1,100	1,100		2,295[2]
Virginia	450	150		
West Virginia	299	299		
Region IV				
Alabama	500	500		500+[2]
Florida	821	821		500+[3]
Georgia	753	628		
Kentucky	450	250		500[2]
Mississippi	319	300		
North Carolina	799	758	85	781
South Carolina	44	42	42	
Tennessee	1,000	755	281[1]	800–900[2]
Region V				
Illinois	224	224	29	1,325[2]
Indiana	1,400	1,400		
Michigan	1,667	1,667	2,019	
Minnesota	117	117	157[1]	300[2]
Ohio	1,000	700	430[1]	1,074[2]
Wisconsin	223	223	60[1]	173[2]
Region VI				
Arkansas	296	108	7[1]	26[2]
Louisiana	499	257		
New Mexico	510	495		
Oklahoma	30	30		
Texas	88	88	29[1]	1,000+[2]
Region VII				
Iowa	370	164	19(37)[1]	384[2]
Kansas	328	314		489[2]
Missouri	1,070	446	54	
Nebraska	40	38		
Region VIII				
Colorado	361	361	159[1]	about 151[2]
Montana	134	132		
North Dakota	47	21		
South Dakota	1	1		
Utah	164	164		56
Wyoming	100	86		
Region IX				
Arizona	503	453	328	about 5,000[2]
California	25,000	6,654		
Hawaii	0	0		
Nevada	0	0		
Region X				
Alaska	164	164		
Idaho	164	164	1[1]	277[2]
Oregon	750	506		
Washington	506	506		700[1,2]

1. Includes some or all NPL sites.
2. Includes unconfirmed sites/potential sites.
3. Investigated/confirmed.
4. Includes all types of hazardous waste sites.

Sources: General Accounting Office. *Survey of States' Cleanups of Non-NPL Hazardous Waste Sites* (Washington, DC: Government Printing Office, 1989). U.S. Environmental Protection Agency, *An Analysis of State Superfund Programs: 50-State Study* (Washington, DC: Office of Emergency and Remedial Response, 1989), pp. 49–52.

The statutes of the states vary considerably. Citizen suit provisions are included in the statutes of 15 states. These provisions allow persons who are or will be adversely affected by a release or threat of a release of a hazardous substance to file a civil action requiring that the responsible parties prevent further damage or take corrective action. Courts may also assess penalties in civil actions filed by citizens. Citizen suits and property transfer programs provide alternative methods for cleanups outside the Superfund process. The objective of a property transfer program is to ensure that real property, in the process of being transferred, does not pose health or environmental risks related to hazardous waste releases. In this procedure owners of certain classes of property must file a negative declaration concerning past or present storage, disposal, or release of hazardous waste at the property, or obtain state approval prior to property transfer. In either case, remediation may be required. Four states—New Jersey, Illinois, Connecticut, and Iowa—have mandatory property transfer programs. Minnesota has a voluntary program.

Provisions for compensating victims of hazardous waste releases exist in 11 states. In 6 of these, the compensation is limited to reimbursement for costs of securing temporary or permanent alternative water supplies. In the other 5, compensation is authorized for a broader array of release-related expenses.

Enforcement authorities contained in statutes that are not specifically intended to address hazardous waste sites are used in 11 states. For example, Michigan relies on enforcement authorities contained in its environmental statutes.

Program Organization

The states' administrative programs for cleanup of hazardous waste sites are centered in state agencies with primary responsibility for environmental matters. The focus of the program may be on the environment and/or health considerations. Many of the cleanup programs are divided into several units. For example, Pennsylvania has a staff of more than 100 people, of which 30 are located in the Department of Environmental Resources (DER) Hazardous Sites Cleanup Program, which has four sections: site assessment, federally funded cleanup, enforcement, and state-funded cleanup. These headquarters' staff are assisted by 42 technical personnel in six regional offices. In addition, the DER's Bureau of Laboratories has 7 persons engaged in state Superfund work, and the Office of Engineering, which is responsible for remedial action contracting, also has 7 positions. The DER's Office of Chief Counsel has 15 lawyers dedicated to the cleanup

TABLE 1.4
State Program Staff Levels

NUMBER OF PERSONNEL	NUMBER OF STATES
More than 100	6
51–100	3
11–50	25
0–10	15
Unknown	1

Source: U.S. Environmental Protection Agency, *An Analysis of State Superfund Programs: 50-State Study* (Washington, DC: Office of Emergency and Remedial Response, 1989), p. 15.

program. Finally, emergency response is handled by a separate program within the DER—each of the six regions has a separate emergency response team of 6 to 12 DER employees.

In a number of states case management teams are used as well as interagency activities that support the cleanup efforts. The staffing levels vary greatly, from more than 600 in New Jersey's Hazardous Waste Management Division to no staff at all in South Dakota and Wyoming (Table 1.4).

Legal Support

State Superfund programs obtain legal support from within their agencies, from the state's attorney general's office, or a combination of these two sources. A total of 23 states report that the state attorney general's office is the sole source of legal support for the cleanup program; within-agency legal personnel provide the only support for 10 state programs; and 16 states rely on a combination of these two for legal support.

When the legal support is divided between the attorney general's office and the agency, the agency legal staff generally provides support on administration and enforcement issues. When a case requires a lawsuit, such as an action for cost recovery, the attorney general's office is normally responsible.

Funding

The availability of a fund or funding mechanism is an essential element of a state's hazardous waste cleanup program (Table 1.5). Typically, a fund is characterized by a depleting and revolving expenditure. A continuing fund allows a state to control the pace of cleanup. If the persons responsible for the waste problem do not cooperate, the state can proceed and then seek punitive damages.

State funds are authorized and/or employed in 48 states for one or more uses relating to mitigation of hazardous waste risks. Fifteen states have more than one fund. The only states without funds are Nebraska and Delaware.

There are 10 types of activities for which funds may be used: remedial activities, CERCLA match, disposal at or development of hazardous waste facilities, emergency response, grants to municipalities and local governments, site investigation, operations and maintenance, removal, studies and design, and victim compensation. Table 1.6 shows the number of funds whose monies are or may be applied to each activity, and the number of states having at least one fund whose monies are or may be applied to each activity.

Emergency responses are the most common activity for which funds have been authorized. Removal of hazardous waste sites is also widely authorized. Studies and design for remedial action are authorized slightly more frequently than remedial action, most likely because of the limited resources of many states to undertake remedial action independently. In 6 of the 11 states, authorizing victim compensation is limited to providing alternative drinking-water supplies (Table 1.6).

A number of the funds have broad provisions for their use. For example, the Pennsylvania Hazardous Cleanup Fund is used to develop recycling and the development of alternative types of land disposal.

TABLE 1.5
State Sources of Funds

| | MAJOR SOURCES | | MINOR SOURCES | |
	Funds	States	Funds	States
Appropriations	20	19	17	17
Fees	20	19	3	3
Bonds	13	12	—	—
Penalties/fines	12	11	29	27
Taxes	10	9	1	1
Cost recovery	6	6	45	41
Transfers	4	4	6	6
Interest	1	1	16	15
General funding	—	—	11	10

Source: U.S. Environmental Protection Agency, *An Analysis of State Superfund Programs: 50-State Study* (Washington, DC: Office of Emergency and Remedial Response, 1989), p. 22.

TABLE 1.6
State Uses of Funds

	NUMBER OF STATES	NUMBER OF FUNDS
Emergency response	47	55
Removals	46	51
Studies and design	42	47
Remedial actions	41	41
CERCLA match	41	47
Operation and maintenance	36	40
Victim compensation	11	11
Site Investigation	8	8
Disposal at or development of hazardous waste facilities	4	4
Grants to municipalities and local governments	3	3

Source: U.S. Environmental Protection Agency, *An Analysis of State Superfund Programs: 50-State Study* (Washington, DC: Office of Emergency and Remedial Response, 1989), p. 24.

Enforcement

Enforcement authorities under the state laws vary significantly (Table 1.7). A total of 37 states have enforcement authorities related to the Superfund programs. Many of these states also use other enforcement authorities in treating hazardous conditions. The major key to enforcement is to reach the parties responsible for the hazardous conditions. The 13 states that rely on non-Superfund authorities cannot always reach the parties that created the hazards. For the most part, state cleanup orders issued under Resource Conservation and Recovery Act–type laws require proof of an RCRA violation. In contrast to CERCLA liability, under these authorities the release of hazardous substances at a site does not support enforcement against former lawful disposers at that site. In some states the state water quality law may provide a basis for enforcement against generators and transporters of hazardous waste.

Of the 37 states with Superfund programs that include enforcement provisions, most have the ability to reach generators and transporters. There is generally no reason for these states to show the existence of a violation of the law. For these states the issue is the scope of liability.

TABLE 1.7
State Cleanup Capabilities and Programs

	CAPABILITIES	PROGRAMS
Region I		
Connecticut	Fund and enforcement	Active cleanup and oversight
Maine	Fund and enforcement	Active cleanup and oversight
Massachusetts	Fund and enforcement	Active cleanup and oversight
New Hampshire	Fund and enforcement	Active cleanup and oversight
Rhode Island	Fund and enforcement	Active cleanup and oversight
Vermont	Fund and enforcement	Fund for cleanup oversight limited
Region II		
New Jersey	Fund and enforcement	Active cleanup and oversight
New York	Fund and enforcement	Active cleanup and oversight
Region III		
Delaware	No fund limited enforcement	1989—State Superfund bill before legislature
Maryland	Fund and enforcement	1988—Allocate funds to state projects
Pennsylvania	Fund and enforcement	1989—Statute authorizes program
Virginia	Fund and enforcement	Funds for cleanup and oversight limited
West Virginia	Limited fund—enforcement only under RCRA-type law	
Region IV		
Alabama	Fund and enforcement	Extremely limited funds—no staff
Florida	Fund and enforcement	Active cleanup and oversight
Georgia	No state Superfund program under Hazardous Waste Management Act	
Kentucky	Fund and enforcement	Fund for cleanup limited
Mississippi	Fund and enforcement	Use other state statutes
North Carolina	Fund and enforcement	Limited funds available
South Carolina	Fund and enforcement	Active cleanup and oversight
Tennessee	Fund and enforcement	Active cleanup and oversight
Region V		
Illinois	Fund and enforcement	Active cleanup and oversight
Indiana	Fund and enforcement	Active cleanup and oversight
Michigan	Fund capabilities; active cleanup	All enforcement from other statutes
Minnesota	Fund and enforcement	Active cleanup and oversight
Ohio	Fund and enforcement	Active cleanup and oversight
Wisconsin	Fund and enforcement	Several statutes—active cleanup
Region VI		
Arkansas	Fund and enforcement	Limited program activities
Louisiana	Fund and enforcement	No funds—oversight continuing
New Mexico	Some fund and enforcement	Limited funds and program
Oklahoma	Some fund and enforcement	Limited funds and program
Texas	Fund and enforcement	Active cleanup and oversight
Region VII		
Iowa	Fund and enforcement	Limited funds for program
Kansas	Fund and enforcement	Active cleanup and oversight
Missouri	Fund and enforcement	Active cleanup and oversight
Nebraska	No funds	Program activity limited
Region VIII		
Colorado	No funds for cleanup	Enforcement under other statutes
Montana	Fund and enforcement	Funds limited
North Dakota	Fund, enforcement in other statutes	Program activity limited
South Dakota	Fund and enforcement	Program activity limited
Utah	Fund and enforcement	Program activity limited
Wyoming	Limited enforcement	Program activity limited
Region IX		
Arizona	Fund and enforcement	Active cleanup and oversight
California	Fund and enforcement	Active cleanup and oversight
Hawaii	Fund and enforcement	Program activity limited
Nevada	Fund and enforcement	Program activity limited
Region X		
Alaska	Fund and enforcement	Program activity limited
Idaho	Limited funds, no enforcement	Program activity limited
Oregon	Fund and enforcement	Active cleanup and oversight
Washington	Fund and enforcement	Active cleanup and oversight

Source: U.S. Environmental Protection Agency. *An Analysis of State Superfund Programs: 50-State Study,* (Washington, DC: Office of Emergency and Remedial Response, 1989), p. 24.

Cleanup Policies

Cleanup policies are key elements in state Superfund programs. A total of 22 states use the Environmental Protection Agency's guidelines of federal standards either as their sole source of cleanup standards or in conjunction with other standards. Standards set by RCRA or CERCLA are important. Of the 22 states using EPA guidelines, 6 use maximum contaminant levels (MCLs) set by the Safe Drinking Water Act as minimum standards for surface and groundwater remediation.

In 18 states, state governments have established their own standards of cleanup for water, soil, and air. Groundwater is of particular concern in a number of states. For example, in California, the potential effect of remedial action on groundwater must be specifically evaluated, and in Minnesota, guidelines for determining site-specific groundwater cleanup goals must be consistent with the state groundwater protection strategy.

Risk standards for carcinogens are used by at least 7 states. A risk standard is applied as an alternative standard to be used when applicable or relevant and appropriate requirements do not exist or as a standard to be achieved at each cleanup. Of the 7 states, Arizona, Indiana, and Minnesota invoke risk standards only in the absence of applicable standards, and California, Virginia, Maine, and Ohio have risk standards that apply generally.

Five states—Florida, Kentucky, Oregon, Pennsylvania, and South Carolina—use ambient quality as the cleanup standard.

Public Participation

The degree of public participation varies greatly from state to state. In 22 states public participation is required under state statute or regulation. In 14 states the state agency seeks public input, while 14 other states do not request public input. In some of these states, public concern will create some participation.

There are 11 states that require public notice at one or more points in the site handling process. Most of these states require notification regarding either site listing or remedial action plans. At least 4 states—Minnesota, Wisconsin, Oregon, and Washington—require notification at several stages during the site handling. In addition, Washington publishes notices of compliance and enforcement orders and of violations.

In 12 states, public comments are solicited on site listings and remedial plans. Of these states, 7 have designated comment periods, ranging from 30 to 60 days. Public meetings or hearings are required by 8 states. In 2 of these—Michigan and Missouri—only an annual meeting is required, either to update a site list or to revise the state program. In 4 states a public meeting must be held upon petition or request. In 12 of the states meetings are held as a matter of policy at the discretion of program officials.

Community relations are a developing feature of state public participation activities. Thus far only Illinois, Louisiana, and Minnesota have developed extensive community relations efforts with regard to hazardous waste sites. In Minnesota each site is assigned a public relations officer, and in Louisiana the Department of Environmental Quality conducts a community relations program at its worst sites. Illinois maintains an active community relations program designed to fine-tune remedy selection using information provided by the local residents. Other states are now drafting community relations programs.

Control of Hazardous and Toxic Waste

The technology available for the control of hazardous and toxic wastes has advanced rapidly in recent years. A number of the more important types of control are described in this section.

Minimization

As the amount of bulk waste increases, so does interest in waste minimization. Every study made in recent years indicates that waste reduction is technically feasible and profitable within existing technology. The goal of waste minimization is the reduction of the amount of hazardous waste generated, so that less must subsequently be treated, stored, or disposed of. Minimization involves the reduction of waste by a generator so as to reduce not only the bulk but also its toxicity, with the goal of minimizing present and future threats to human health and the environment.

Waste Reduction

There are a wide variety of ways to reduce the bulk of hazardous wastes. In the industrial process, reduction measures include process modifications, feedstock substitutions, improvements in feedstock purity, implementation of housekeeping and management procedures, increases in the efficiency of machinery, use or reclamation within a process, and any other action that reduces the amount of waste leaving a process. Many hazardous wastes have useful products associated with them. Reclamation is the treatment to remove the usable product and at the same time reduce the bulk of hazardous materials. In some processes it is also possible to include the direct and effective substitution of waste for an ingredient or raw material used in the industrial process or for a commercial chemical product. Treatment that is intended solely to prepare the hazardous waste for disposal, such as dewatering, is normally not considered waste minimization.

Engineers who have studied waste reduction implementation have concluded that a critical initial step is to conduct a *waste reduction assessment* (WRA). The steps in this process are as follows:

1. Identify the amounts and kinds of hazardous substances in wastes.
2. Identify the specific production sources of the wastes.
3. Establish priorities for waste reduction based on such criteria as costs, environmental benefits, health and safety liabilities, and production restraints.
4. Select technology that is economically feasible.
5. Compare the economics of waste reduction alternatives with current and future waste management control options.
6. Evaluate the progress and success of chosen waste reduction measures.

If waste reduction is to achieve its full potential, the waste generator must consider both the technological aspects of waste reduction and the human aspects. Among the issues to be weighed are economic impacts, the reduction in health hazard, and the influence on the environment. This requires a major effort in research and development.

Benefits

The basic purpose of minimization of hazardous wastes is to reduce health and environmental risks and at the same time reduce costs. The strategy of the modern manager is to maximize the interaction between technology and the use of a resource. Thus a nonwaste technology must be developed in which resources are conserved and pollution reduced.

In addition to the reduction of waste itself, there are other benefits to manufacturers from the minimization of hazardous waste, such as reduced costs of transporting, treating, storing, and/or disposing of the waste. Savings may result also from the reduction of long-term liability that should come with decreased handling of wastes, although these are difficult to quantify; savings may also be realized as regulatory costs of waste in general decrease or if regulated treatment or storage facilities are eliminated. Another benefit companies may reap is the development of goodwill with the neighborhoods in which they operate.

Progress

The benefits of waste reduction were recognized in theory about 1980, but practice has been slow to follow, although technology exists. There has been little interest at the federal regulatory level in encouraging waste minimization. Environmental organizations have also been relatively inactive in advocating waste reduction because of worry about loss of support for established regulatory programs. Private industry has also not supported a federal waste reduction program because of concern about possible secondary impacts. Industry fears burdensome regulation of waste reduction. Companies involved in waste management and pollution control fear loss of business. These concerns have impeded the development of federal policies required to develop strong waste reduction efforts.

Incineration

The incineration of hazardous waste as a solution to processing and disposal has long been a controversial subject. Viewpoints range widely; some see incineration as the ultimate answer, while others see it as one of the most devastating and dangerous practices perpetrated by engineers. Major advances have been made in the past decades that provide insight into the limitations and advantages of incineration. In general, however, the public remains skeptical because of fear of neighborhood air contamination. This has created a political environment in which it is difficult to obtain a permit for, construct, and operate an incinerator.

Characteristics

The definition of incinerable waste has been broadened to include wastes that are not capable of sustaining combustion. Wastes that are now considered appropriate for incineration include any that are hazardous when they contain toxic organic compounds, regardless of the toxic concentrates or the matrix in which they occur. Even contaminated water and soil, as well as many other solids and liquids that contain trace levels of organic compounds, are now routinely being incinerated.

There are now only two classes of hazardous wastes that are not considered appropriate for incineration. The first is any hazardous material that contains toxic heavy metals. Incineration cannot destroy heavy metals; although burning reduces the toxicity, it cannot remove the metals themselves. The heavy metals remain in the bottom of the incinerator or in the fly ash.

The other class of wastes that cannot be incinerated are those with a high halogen content. Wastes with a chlorine and/or fluorine content in excess of about 30 percent are difficult to incinerate because of the formation of highly corrosive gases. These gases can damage the incinerator.

Principles of Incineration

The basic concept of incineration is that it is a precisely engineered and controlled process of destroying organic materials by means of high-temperature thermal oxidation. In this process all organic waste material is converted to carbon dioxide and water. When properly designed and operated, incinerators of solid and liquid wastes have an efficiency rate of at least 99.99 percent.

In the development of a satisfactory incineration process, the three factors of time, temperature, and turbulence are critical. *Time* here refers to the amount of time that a substance needs to be exposed to high temperatures in the incinerator in order to destroy its toxicity. This is known as residence time. The longer the exposure, the greater the destruction of the waste molecules. Most modern incinerators are designed with residence times of more than 2.0 seconds.

Temperature refers to the amount of heat required to destroy organic substances. Most modern incinerators are designed to operate at temperatures in excess of 2,000 Fahrenheit. As with time, the higher the temperature, the more effective the process.

Turbulence refers to the amount of mixing of the waste in the incinerator with the hot gases around it. The greater the turbulence, the greater the heat transfer from the incinerator flame to the waste materials. This is a critical consideration in the destructive process. A well-constructed incinerator system achieves a balance among these three factors.

Incinerator Technology

A number of types of incinerators have been devised, each of which has a particular characteristic. The most common type of incinerator found in commercial hazardous waste incineration is the rotary kiln incinerator, because it is extremely versatile. This type of unit can burn tars, organic and aqueous liquids, sludges, homogeneous granular solids, and irregular bulk solids. Rotary kilns can also burn waste with a high fusible ash content. The average temperature of a rotary kiln incinerator is about 60 million Btus per hour, but can be as large as 90 million Btus per hour.

Fixed-hearth incinerators are primarily designed for the combustion of homogeneous granular solids, irregular bulk solids, and heavy tars, but they can be designed for liquid, organic, and aqueous waste materials. The fluoridized bed incinerators offer somewhat more flexibility than fixed-hearth incinerators in terms of waste materials that can be burned. These incinerators can be used to destroy homogeneous granular solids, heavy tars, aqueous and organic sludges, and aqueous and organic liquids. They are not designed to burn irregular bulky solids. Liquid waste incinerators are common in the petroleum and chemical industry for on-site destruction of liquid waste materials generated at these plants. These types of incinerators can be used to incinerate any type of solid or sludge waste. Other types of incinerators include cyclonic, auger, infrared, and plasma.

Regulatory Requirements

Hazardous waste incinerators are regulated by the Resource Conservation and Recovery Act, except when they are used to burn polychlorinated biphenyls (PCBs). The incineration of PCBs is controlled by the Toxic Substance Control Act. These regulations require that all wastes disposed of by incineration must be destroyed with an efficiency of at least 99.99 percent (99.9999 percent for PCBs). This efficiency must be demonstrated by an extensive trial burn test program conducted by the Environmental Protection Agency.

Other requirements that are placed on hazardous waste incinerators pertain to emissions of particulates and hydrogen chloride. The particulate matter concentration in emissions cannot be greater than 0.08 grams of particulate matter per dry standard cubic foot of exhaust with the exhaust corrected to a level of 7.0 percent oxygen. Some states have more stringent requirements. Federal regulations limit the amount of hydrogen chloride that can be emitted to 4.0 pounds per hour or, if over 4.0 pounds per hour, no more than 1.0 percent of the amount of hydrogen chloride present in the exhaust prior to the emission control equipment. There is also a requirement of continuous monitoring by incinerator operators.

Landfills

One of the traditional ways to deposit waste in the United States has been in landfills. The technique developed as a place to store new sanitary waste. However, over the years, contamination of landfills by hazardous wastes appears to have become widespread. In the Midwest it has been estimated that 90 percent of the municipally operated landfills contain hazardous and toxic wastes. In many areas, there is a recognition that landfills are no longer safe depositories and that waste is polluting the environment.

The U.S. Superfund program has targeted for study about 140 former municipal landfills in 27 states, and numerous others are being studied by state agencies. Prior to July 15, 1985, "small generators" producing less than 2,200 pounds per month of hazardous waste were not required to document where wastes were deposited in licensed disposal facilities. Under new regulations, industries generating more than 220 pounds per month (about half a drum) must report the type of disposal of such wastes. Officials fear that because the firms are small and have little capital, they may illegally dispose of their wastes in municipal landfills to avoid paperwork and disposal fees of $50 to $250 or more per drum.

The hazardous waste problem in landfills exists as a result of lack of controls in the past. Undoubtedly, thousands of municipalities have accepted industrial wastes in their landfills as a free or at-cost service to local manufacturers. This practice often prevailed when segregated landfill space was scarce, or when industries paid high "fees" at special "service cell" facilities. The hazardous wastes thus disposed of have included cyanide, heavy metals, synthetic organic chemicals such as chlorinated solvents, PCBs, pesticides, and other petroleum derivatives from a variety of industries. These practices violated no laws.

As environmental controls have evolved, new laws have been enacted to control hazardous wastes in landfills. The Resource Conservation and Recovery Act of 1976 contains provisions for the management of solid and hazardous wastes. The Comprehensive Environmental Response, Compensation and Liability Act of 1980 has addressed this problem. The transportation of hazardous wastes, including PCBs, is also regulated under RCRA in accordance with the U.S. Department of Transportation labeling, packing, and shipping regulations.

When a municipally owned or operated facility is found to contain hazardous waste, there are three pathways to removal of the problem. If the landfill was closed before November 19, 1980, it falls under the regulation of the Comprehensive Environmental Response, Compensation and Liability Act of 1980. Responsible parties must eliminate the waste so that environmental contamination is prevented.

If the landfill accepted hazardous wastes after November 19, 1980, it falls under the regulation of the Resource Conservation and Recovery Act, and the Environmental Protection Agency has the responsibility for its cleanup. If the landfill does not notify the EPA of the environmental problem, the landfill operator can be subject to RCRA and/or CERCLA penalties. In addition, since the 1984 Hazardous and Solid Waste Amendments to RCRA, solid waste management teams must provide corrective action to contain and control pollutant releases.

The Environmental Protection Agency makes a strong effort to determine who deposited the hazardous materials and, in the cleanup, to recover remediation costs. In addition to the waste generators, the municipality normally has to bear some of the cost of remediation. Cleanup and control are very costly, and the cost increases if an aquifer is contaminated.

There is some evidence that the disposal of hazardous waste in landfills is increasing. This creates some difficult problems for local governments. They must remain abreast of unanticipated regulatory actions and continue to provide quality waste management services. Of the current regulations, RCRA will be most challenging to local authorities, for it requires compliance with a host of regulations in addition to corrective action provisions.

An RCRA-regulated landfill originally operated in one of two ways, known as interim status or final status. Under interim status a facility had to submit a narrative description of facility operations and locations. From 1985 onward, all facilities had to secure permits, as the interim status was eliminated. The penalty system can be complex, expensive, and time-consuming, and involves substantial liability issues. These factors require a careful assessment by the local community.

A number of alternative options have been created to the acquisition of a landfill permit. An exemption may be requested. This occurs usually in facilities handling a single waste stream that is readily detoxified, such as by adding lime to the landfill waste. A number of surface impoundments in Utah, Ohio, and Pennsylvania have been diluted through this method. It is not likely that a landfill with a mixture of toxic wastes buried indiscriminately in it could acquire an exemption.

Landfills that accepted waste prior to January 28, 1983, may qualify for closure of the site under the interim status regulation. For example, if toxic substances are not leaking into groundwater, the need for monitoring is decreased. Depending upon site-specific conditions, closure of a landfill may not be permitted.

Another option is to close the landfill with the waste on-site, but stabilized so that the toxic substances become inert or nontoxic. In order to secure the area from waste contamination, an impermeable surface cap and run-on/run-off control are usually needed, for the migration of water must be controlled. Closure is not permitted in some states. State agencies may require maintenance and monitoring of groundwater, either through RCRA or through state solid waste regulations.

An additional alternative is the possibility of removal of all regulated waste to a licensed RCRA facility. To meet acceptable standards, landfill operators must demonstrate that hazardous waste no longer exists at the site at levels above background or above the standards acceptable to federal and state agencies. This alternative has been largely limited to storage or treatment of surface impoundments with relatively little subsurface migration.

As environmental awareness has grown, the investigation of the potential of hazardous material in landfills has also increased. In the future, many municipal landfill facilities will be confronted with the problem of disposal of hazardous wastes. Unless adequate state or federal funding is provided, or reliable private companies identified, municipalities are likely to be responsible for some portion of the financial and environmental cleanup. The new laws and regulations are complex and rigorous, and proper compliance, particularly under RCRA, requires considerable expertise in a number of technical areas. The regulatory process that governs the control of hazardous wastes in landfills is technical and involves a considerable potential for long-term liability.

Biological Treatment

In biological waste treatment, the individual microbial cells can be considered small chemical reactors capable of converting a waste compound to a less harmful substance. Their conversion results in a change of chemical energy in the system of reactants and products. In essence, in biological treatment processes wastes are gradually broken down through a series of chemical reactions. The free energy that is released is used in part by the microorganisms for their metabolic needs.

The technology for microbial distribution of wastes had its origin in the activated sludge process for sewage treatment developed around the turn of the twentieth century. As industry developed, the chemical structure of the wastes to be treated became increasingly less similar to those found in nature, and other techniques were also developed. Nonetheless, the microbial process has continued to be effective due to the remarkable ability of life to adjust to environmental changes. In general, the microbial organisms have been able to assimilate most organic compounds produced by industry, although sometimes only slowly and at low concentrations. Microorganisms have also demonstrated the capability of treating inorganic matter by either direct or indirect metabolic processes. As a consequence, microorganisms are utilized to metabolize many different compounds in domestic or industrial wastes that are toxic, carcinogenic, or otherwise generally undesirable in the environment. Scientific advances have been rapid, and we now have the capability to assist nature with the process of adaptation by selectively changing small facets of microorganisms to improve their effectiveness in treating pollutants.

The speed of microbial conversion depends upon a number of factors. If the reaction is slow to proceed, it is because the reactants must first overcome an activation barrier before they are changed to products. Increasing the temperature of the reactants is one way to overcome the activation energy barrier. One technique is to lower the activation energy barrier through the use of a chemical agent, known as a catalyst. It is through this procedure that biological activity is most effective. The biochemical reactions within the microorganisms produce catalysts called enzymes that increase the waste decomposition reactions.

Enzymes consist of both proteins and nonproteins. The protein portion of an enzyme, called the apoenzyme, is considered to be responsible for the chemical specificity of catalytic action. The nonprotein portion of the enzyme is called the pro-factor. Pro-factors are considered to be responsible for affecting the chemical reactions, such as oxidation and hydrolysis transfer.

Microbial technologies have proven particularly attractive for treating solid industrial waste, because of their ability to process wastes in situ. In addition, such treatment may eliminate the need for large capital expenditures at processing plants, such as incinerators. The technique has been applied successfully in a number of industries for more than half a century. These include the petroleum industry, where the microbial process has been used to treat oily spills by controlled application in water or topsoil when there is a rich microbial organism population. The extension of these techniques to organic chemicals such as chlorinated biphenyls, pesticide residues, and dioxin may be possible with additional research.

A large number of research programs aimed at expanding the use of biological technology are currently in progress. Opportunities for process improvement exist in both the engineering and the biological aspects of the technology. The engineering improvements will come quickly. In contrast, the biological improvements—that is, the identification of the best types of organisms to attack a particular waste—are likely to require longer research and development. The impact of genetic engineering could be substantial. However, legal precedents for the use of recombinant mutant organisms in open environments have not been established, and extensive political, emotional, and technical questions regarding environmental safety must be determined by the courts and the public. While the problems are great in genetic engineering, it is a rapidly developing area for the treatment of hazardous wastes.

Chemical Treatment

When it is desirable to alter the nature of materials, chemical treatment is a widely accepted technique for the disposition of hazardous wastes. To illustrate, it is possible to convert the pesticide ethylene dibromide (EDB) chemically to a salt, potassium bromide (which has economic value), and the gas acetylene, or for glycol-based chemical reagents to remove a chlorine atom from 2,3,7,8-dioxin, converting it to a nontoxic state. Of the chemical treatment processes some of the most important are oxidation, reduction, neutralization, chlorinolysis, and dechlorination.

In the oxidation process ions or compounds can be chemically oxidized to render them nonhazardous. In the process, the chemical oxidizing agent is altered. This process has a number of applications in treating certain types of hazardous wastes. One of its major applications is the treating of organic and inorganic contaminants in aqueous solutions. For example, a major use has been for removing cyanide from metal-plating wastes. Because oxidation is relatively nonselective in oxidizing contaminants, the process has limited application to slurries and sludges, which may contain many oxidizable components. It is most useful in diluting aqueous wastes.

Ozone may be used in the oxidation process. It is particularly applicable for aqueous streams that contain less than 1 percent oxidizable compounds. Ozone is effective in breaking down refractory organisms in waste before they are subjected to biological or other treatment. A combination of ultraviolet radiation and ozone is effective in treating such halogenated organics as aldrin, dieldrin, endrin, kepone, and chlorinated phenols.

Chemical reduction involves the transfer of electrons from one compound to another and is used either to make compounds nontoxic or to enable compounds to undergo chemical change or physical removal. Chemical reduction is most effective when applied to liquid wastes free of organic compounds. For example, it is widely used in industry to control chromium waste, to remove mercury from mercury cells, and to reduce complex metals, such as copper and nickel, by precipitation out of solution.

Neutralization, because of its ease in use, is one of the most common and practical processes used in hazardous waste treatment. It involves combining a hazardous waste with either an acid or a base to produce a liquid with a pH reading at acceptable levels. Neutralization is frequently used before a final waste treatment in order to protect the immediate ecosystem. Neutralization normally occurs in corrosion resistant tanks, although it may occur in ponds or limestone filter beds. Bases commonly used in neutralization include lime, calcium, hydroxide, caustic soda, soda ash, and ammonium hydroxide. The acids used include sulfuric, hydrochloric, and nitric.

In the chlorinolysis process, chlorine reacts with chlorinated hydrocarbon waste at temperatures of 500 to 800C. Generally, this process is suitable for treating chlorinated wastes and residues, including pesticides, herbicides, solvents, and vinyl chloride. The stages in the chlorinolysis process include (1) pretreatment of the waste to remove water, solid materials, and compounds extraneous to the process; (2) reaction with chlorine; (3) distillation to separate reaction products; and, finally, (4) the absorption of the hydrochloric acid solution that is produced.

Dechlorination uses chemical reagents to remove chlorine from chlorinated molecules, to break up chlorinated molecules, and to change the molecular structure of the molecule. Metallic sodium is a typical reagent used to remove chlorine from substances to form sodium chloride. Most of the research has been aimed at the detoxification of PCBs, but the process is applicable to removing many chlorinated organic molecules. For example, dioxin and PCBs can be removed from soils using a series of reagents prepared from potassium and polyethylene glycols.

Physical Treatment

A wide range of physical treatment processes are used in handling hazardous waste. In the process of separating solids from liquids, sedimentation and filtration techniques are commonly used. Sedimentation basins are frequently used in the removal of solids from water. The separation of chemically coagulated materials provides an economic method of liquid/solid separation. Filtration can be accomplished with microscreens, diatomaceous earth filters, sand filters, and mixed media filters. For example, the filtration of sludges is accomplished with sand beds or vacuum filters.

Distillation and evaporation are procedures employed to separate waste streams into two or more factions. A number of techniques have been developed. To illustrate, in a stripping operation air is forced through the waste in order to collect vaporized materials. Air stripping has been used to remove trace concentrations of volatile solvents from contaminated groundwater. In waste oil recovery the lighter hydrocarbons can be evaporated from the heavier oil wastes and recovered.

Absorption or activated carbon also permits concentration of organics dissolved in water. In many hazardous waste sites suspended solids and a wide variety of organic materials are allowed to settle as a sludge and are disposed of separately. The concentration of a wide variety of organic chemicals is thus possible.

Stabilization

Stabilization is a technique that reduces the hazard potential of wastes by converting them into their least soluble or mobile form. In stabilization processes the wastes are encapsulated in a solid of high structural integrity. The solidification does not necessarily change the physical or chemical nature of the waste, but may involve the mechanical binding of the waste into the solid. The migration of toxic material is greatly retarded by the decreased surface area exposed to leaching.

Most current stabilization systems are proprietary processes, involving the addition of absorbents and solidifying agents to wastes. Most of the processes fall within a few generic types—sorption, lime- and/or cement-based techniques, thermoplastic microencapsulation, and vitrification.

The sorption technique involves the adding of a dry, solid substance to liquid wastes to absorb the free liquid and to facilitate the handling characteristics. The sorbent may hold the liquid in capillary form or may react chemically with it. Common natural sorbents are soil, fly ash, cement kiln dust, and lime kiln dust. Synthetic sorbents are also available. Sorbents are widely used at hazardous waste landfills, but sorbents that soak up only liquids are prohibited by the RCRA amendments of 1984.

The lime- and cement-based techniques involve the use of hydrated lime and natural or artificial silicalike material such as pozzuolana. Natural pozzuolana includes blast furnace slag and diatomaceous earth. Artificial pozzuolana includes furnace slag, ground brick, and some fly ashes. Lime and pozzuolana are frequently mixed together. A number of processes use portland cement as a solidifying agent. In this technique the suspended solids in the waste slurry are incorporated into the hardened matrix. For example, most toxic metals are transformed into the low-solubility hydroxides or carbonates by the high pH of the cement mixture. Some materials, such as sulfides, asbestos, and latex, can increase the strength and durability of cemented wastes.

Thermoplastic microencapsulation involves the mixing of dried wastes with a heated thermoplastic material such as asphalt, paraffin, or polyethylene. The most commonly used of these agents is asphalt. The technique began with the disposal of radioactive wastes but has been utilized for the control of highly soluble toxic substances, which are not amenable to lime or cement techniques. The technique is costly and is normally employed when complete containment of special hazardous waste is required.

Microencapsulation is commonly referred to as jacketing. This technique surrounds the waste product with a durable, impermeable coating. An established technique involves sealing the waste in a drum made of or lined with polyethylene.

Vitrification is another solidification technique. In this process, wastes are mixed with silica, heated to extremely high temperatures so that the molecules melt, and then allowed to cool into a glasslike solid. It is now possible to vitrify buried wastes in situ by using graphite electrodes. In this process there is a danger of toxic gases being released, so that hoods are essential to capture these gases. This method is energy intensive and costly, so it generally is used only for radioactive or extremely dangerous wastes.

Underground Burial

As hazardous wastes increased in the 1960s and 1970s, the practice of burying waste products at shallow depths developed. This type of disposal too often was simply concealment rather than true isolation from human contact. As a result, many contaminants that jeopardized the total ecosystem entered the biosphere. This indirect effect upon humans usually did not become apparent for months or years, making the source of the problem more difficult to determine and corrective measures vastly more complex and costly.

During the 1970s, growing public awareness and increasingly strict federal regulations encouraged the development of new burial techniques, such as well injection, deep mines, and natural and human-made caverns. It has been determined that the most fundamental requirements for successful burial of waste center upon effective control of surface and subsurface water. Such a burial site must be characterized by the following:

1. Sufficient depth to accommodate the buried waste
2. A water regime so that contaminants will not migrate throughout the area

3. Soil or bedrock sufficiently homogeneous and impermeable to ensure slow, predictable groundwater movement away from the site

4. A sufficiently stable geological environment to ensure that conditions suitable for waste disposal at the time of burial will not be substantially altered during the anticipated useful life of the site

5. Suitable slope, seismic activity, lithology, climate, and other significant factors in site performance, primarily as they affect one or more of the above essential properties

The complex problem of burial waste site selection has been a subject of extensive study in recent years, not only for hazardous and toxic materials but also for radioactive wastes. The rocks that make up the earth's crust, though varied in origin and composition, have developed over periods spanning thousands to millions of years. Obviously, certain areas are not suitable for storing hazardous materials. Areas of rapid tectonic change when earthquakes or volcanic eruptions occur are unsuitable. Areas where rapid erosion is taking place or areas that are subject to floods are equally unsuitable. Most of these areas can be readily identified.

In the evolution of a site for hazardous material a number of geological selection schemes have been devised. The following procedure presumes that waste containment will be achieved through the geological characteristics of the site, rather than through the use of engineered containers or structures. Using the following steps, one or two optimum sites may be identified in a region of not more than 50,000 square miles. This procedure is based on the concept that some form of low-cost rapid regional screening program is essential because a detailed survey of a large area is prohibitively expensive, in both time and money.

The following four levels of evaluation are required in evaluating an underground waste disposal site.

1. General Exclusionary Criteria. At this initial stage the regional characteristics are evaluated which would most severely jeopardize site performance. This includes evaluation of such physical factors as volcanic activity, seismically active region, coastal zones and flood plains, fracture or cavernous bedrock such as limestone because of rapid and often unpredictable movement of groundwater through such rocks. In addition, hazardous waste sites must be located at a sufficient distance from population centers and public water supplies to afford reasonable protection in the event of any form of failure.

2. Site-Specific Exclusionary Characteristics. The physical details of the area are investigated at this stage. Such aspects as minor earthquakes, stream erosion, landslides and other local conditions that may endanger a hazardous waste site.

3. Interactive Site-Specific Performance Characteristics. Many geological factors of a potential site are related to other factors such as site size, depth of water table, soil characteristics that are important in predicting site performance. For example, if high soil transmissivity exists, then the size of the site must be increased to ensure that contaminated leachate cannot escape from the controlled area. Further, many characteristics of a particular site may afford economic, operational, or socio-political benefits which will influence site selection. Such features as a transportation network and bedrock composition which may absorb or neutralize some of the waste materials may affect the final site selection. (W. F. Witzig, W. P. Dornsife, and F. A. Clemente, eds. *Low Level Radioactive Waste Disposal Siting: A Social and Technical Plan for Pennsylvania* [4 vols.]. Chapter 6. University Park: Institute for Research on Land and Water Resources, The Pennsylvania State University, 1983)

As shown above, an effective underground disposal site for hazardous wastes must exhibit a complex and critical association of interrelated environmental and social characteristics. Although such sites may not be rare, they are difficult and costly to identify. An underground waste storage site must be considered a national resource in the same way minerals or natural beauty are considered a part of national heritage.

Environmental Toxic Chemicals

Toxic chemicals are substances that are harmful to living organisms. There are literally thousands of chemicals that are toxic to animals, humans, or both. Some of these chemicals occur naturally; others are human-made. Many toxic chemicals can be measured in the environment at background levels, but are not harmful to the environment or to human health.

Many manufactured chemicals are deliberately released into the environment in order to control particular problems. Pesticides, released for the control of insects and weeds, are an obvious example. Other toxic chemicals enter the environment as waste from industrial processes. These include deliberate discharges and those that result from leaks or accidental spills.

Toxic materials may be released by any of three media—air, water, or land. Once released, they may move and mix and react in the environment. They may be transported great distances by a number of means. The toxins may move from land surface into groundwater or from any medium to another through deposition or volatilization. In the process, they may be absorbed and concentrated in living organisms through food chains. The magnitude of these effects varies according to the characteristics of the chemical and the environment. If a toxic chemical is to be controlled, understanding the way it moves in the environment is essential.

As toxic chemicals move through the environment they create risks of exposure. Exposure may occur in the form of a single major dose of a single chemical, as in a chemical accident, or cumulatively, as in an individual's exposure to complex mixtures over a long period of time. Exposure to toxic chemicals from a single source or in a particular environment must be assessed in the context of the individual's total exposure.

Sources of Environmental Toxic Chemicals

As noted, toxic chemicals occur both naturally in the environment and through manufacture, including by-products of industrial processes.

Natural Sources

Nature produces many toxins. One that is common is aflatoxin, found in corn, rice, sorghum, peanuts, and many other foods. It is the product of a fungus, *Aspergillus flavus*. It occurs worldwide, but is most common in wet, warm environments. Many plants also produce toxins of their own. For example, the toxicity of some mushrooms has been recognized for centuries. Plants such as jimsonweed, hemlock, and oleander contain alkaloids that can cause acute illness or death.

Manufactured Sources

Today there are more than 62,000 chemicals manufactured in the United States. Many of these chemicals are hazardous to human health. As knowledge about the effects of chemical pollutants has grown, there has risen the need for greater understanding of these chemicals on human health. It was not until the 1980s that accurate data began to be collected for predicting dangers to humans and for estimating how human exposure might change in response to different regulatory policies.

Toxic Chemicals and Human Interaction

There are at least hundreds, and more likely thousands, of toxic chemicals that have very restricted applications. The uses of these particular toxic chemicals can easily be controlled. Other toxic chemicals have wider applications, however, and major problems arise when these are widely distributed through the environment in the attempt to destroy one undesirable element. The use of pesticide sprays is the best example of such use; pesticides are often sprayed on hundreds of thousands of square miles of land. Many of these toxic chemicals are long-lived, and they enter the food chain, ultimately reaching human consumption.

Although much research has been conducted on the effects of toxic chemicals on human health, there are still many unanswered questions. Long-range studies are required to determine the variations among humans of threshold levels for toxic chemicals. Many decisions are currently being made based on short-range studies. In spite of the best possible information, all efforts to control toxic chemicals will fail if (1) industry puts profits ahead of safety, (2) governments do not develop institutional plans for human welfare, and (3) people lack the will to develop policies to ensure safety in the use of toxic chemicals.

The following examples illustrate some of the most toxic widely used chemicals, and attempts to control their use.

PCBs

PCBs are a family of at least 209 chemical compounds consisting of two benzene rings and two or more chlorine atoms. They vary in character from light, oily fluids to heavy, greasy or waxy substances. Although they were discovered more than a hundred years ago, commercial production began only in 1929, when the remarkable insulating capacity and nonflammable nature of these compounds were recognized. PCBs replaced combustible insulating fluids and therefore reduced the risk of fire. They were so effective that some cities banned the use of other types of mineral oils. For a number of decades PCBs were routinely used in the manufacture of such products as plastics, adhesives, paints, newsprint, and caulking compounds. Between 1929 and 1979 it is estimated that about 1.2 billion pounds of PCBs were produced in the United States.

It had long been known that PCBs were among the most poisonous chemicals ever produced. In the 1960s the health effects of small amounts of PCBs in the human body were uncertain, but large exposures had been associated with miscarriages, birth defects, liver disorders, hypertension, and skin rash. A major poisoning incident occurred in Japan in 1968, when more than 1,600 people became ill after consuming rice oil contaminated with PCBs. Victims developed skin disorders, nausea, and swelling of arms and legs; some developed liver disorders.

The growing concern about PCBs as a danger to health and environment was attributed to how PCB wastes were handled. Industries manufacturing and consuming PCBs were usually permitted to discharge PCB-laden water into rivers and streams or to deposit it in open landfills. While these practices were usually legal, they were potentially harmful to the environment.

These waste disposal practices resulted in the accumulation of PCBs in the environment, for PCBs decompose slowly because they are relatively insoluble in water. However, they are highly soluble in fats, and if they enter the food chain they remain as residues in the body of humans. Tests on animals have raised a number of questions about the possible health hazards of PCBs. A number of studies suggest that PCBs may cause an increase in liver cancer. Other studies, however, question whether PCBs are carcinogenic. In any case, in 1977 the Environmental Protection Agency concluded that PCBs are a potential health hazard and stopped their production in the United States. It is estimated that about 750 million pounds are still in use in this country. Italy, France, Spain, and a number of other European countries continue to manufacture PCBs, although little is imported into the United States.

Alternative materials such as silicone fluids, fluorocarbons, and mineral oils have replaced PCBs in manufactured products. All of these substances are less effective than PCBs. For example, a transformer cannot operate at the same power load with a substitute chemical. In addition, many of these substitutes present fire or other environmental risks.

Many efforts are being made to clean up the PCB-polluted environment. The removal of PCBs from landfills is an ongoing process. Cleanup of contaminated rivers is also under way. For example, from 1947 to 1977 more than 500,000 pounds of PCBs were discharged, under permit, into the Hudson River from two General Electric plants at Fort Edward and Hudson Falls. "Hot spots," which contain 50 ppm (parts per million) or more of PCBs have been identified. The PCB concentrations range from 5 to 1,000 ppm in fine-grained sediments. Millions of dollars have been spent to remove the PCBs. It is estimated that between the efforts of the cleanup and the natural discharge of sediments from the river, the Hudson will be free of PCBs sometime between 2001 and 2013.

There is every indication that the problem of the contamination of the environment by PCB wastes could have been avoided if it had been recognized that a potential health hazard existed. Between 1947 and 1975 there was no effort to keep PCBs out of the environment. A major lesson can be learned from this unfortunate case. The use of a valuable product has been lost through ineptitude and shortsightedness, because PCB wastes were deposited in the environment without a thought about the possible consequences.

Lindane

While most pesticides are important to control insects, pests, and fungi in natural surroundings, lindane was the most widely used chemical indoors in the home. It grew in use despite its extreme toxicity, which has been recognized since around 1900. Since the 1970s lindane has been suspected of causing cancer. Lindane has many uses; it has been employed not only to kill insects directly, but has been used in impregnated shelf paper, in floor wax, for dog shampoo, and for treatment of small grains. Similar to other chlorinated hydrocarbons, lindane is absorbed through the skin.

Because of adverse health reports the Environmental Protection Agency began a formal review of the chemical in May 1976. In October 1980, the EPA's Scientific Advisory Panel completed its review. The report indicated that the use of lindane should be stopped immediately in such items as home shelf paper and floor wax, pet products, and ornamental applications in the home. The report recommended that lindane continue to be used, however, by commercial nursery workers, in seed treatments, and for selected crops, livestock, pest control, and other like uses. Environmentalists continue to press for greater control of lindane. It remains one of the nation's widely used pesticides.

Chlordane and Heptachlor

Chlordane and heptachlor, along with DDT, were among the first widely used pesticides to receive national attention. As far back as 1950, Dr. Arnold Lehman, chief pharmacologist of the federal Food and Drug Administration, stated that chlordane was "one of the most toxic of insecticides—anyone handling it could be poisoned" (Rachel Carson, *Silent Spring*, 1962).

Chlordane and heptachlor were used to control insect pests in agriculture and in the home. The spraying of these pesticides has a particularly negative effect on the wildlife population of the region. There have also been numerous reports of humans exposed to chlordane and heptachlor developing leukemia and other disorders. These chemicals can enter the body in a variety of ways, through ingestion of contaminated food, inhalation in the form of dust or spray, or absorption through the skin.

In 1971 the Environmental Protection Agency began a review of animal tissue slides that the agency had accepted to demonstrate heptachlor's safety. In 1974 the same tests were carried out for chlordane. In both tests it was shown that both chemicals caused cancer ("The Environmental Protection Agency and the Regulation of Pesticides," Staff Report to the Subcommittee on Administrative Practice and Procedure, U.S. Senate, December 1976, p. 24). To contest these reports, Velsicol Chemical Corporation in 1975 submitted a report to the EPA stating that no tumors had been found in test animals. This proved to be a misrepresentation and in December 1977 several Velsicol officials were indicted by a federal grand jury in Chicago and charged with concealing from the EPA studies showing that chlordane and heptachlor can cause cancer. Due to irregularities in the grand jury investigation the charges were dropped. The case was resolved in 1981 when Velsicol, facing a possible fine of $1 million, was fined only $1,000 by a U.S. District Court in Chicago.

As a response to the adverse health reports, EPA Administrator Russell E. Train in November 1974 issued a "Notice of Intent" to cancel the use of chlordane and heptachlor. Velsicol protested the cancellation proposal. Evidence continued to build against the insecticides, and in December 1975 Train announced an immediate temporary ban on most applications of the chemicals, suspending their use on lawns, gardens, turf, and for household pest control. The chemicals could still be used on corn, strawberries, pineapples, citrus fruits, termites, and other crops and insects. The battle over the use of the chemicals continued by the environmentalists and EPA. Finally, in March 1978 the EPA agreed to phase out most uses of chlordane by 1980 and heptachlor by July 1983.

Toxaphene

Toxaphene, a chlorinated hydrocarbon, was at one time the nation's most heavily used insecticide, accounting for one-fifth of all pesticides in the United States. It was first used in 1947, mainly on such crops as cotton, rice, cranberries, grains, and vegetables to kill parasites. At the height of its use 100 million pounds were consumed each year. One of toxaphene's greatest dangers is its permanence. The National Cancer Institute (NCI) has stated that it "appears readily transported from its site of application either by water or by air. . . . It is persistent in soil and water and accumulates at increased concentrations in aquatic life." Thus it enters the food chain and ultimately reaches human beings.

The Environmental Protection Agency began its review of toxaphene in the early 1970s. In 1977, when the EPA released its initial review, it cited a 1976 NCI study revealing that "a significant increase in the incidence of cancerous growths did in fact develop" in laboratory animals fed toxaphene (*Federal Register*, May 25, 1977). Dr. Adrian Gross of the EPA, formerly associate director of the Food and Drug Administration, reported, it is "abundantly clear that toxaphene is an extremely potent carcinogen in rats as well as mice" ("Toxaphene: Position Document 1," Toxaphene Working Group, U.S. Environmental Protection Agency, April 19, 1977).

Toxaphene has had its most adverse effects on wildlife and cattle. The EPA reports that toxaphene was a major factor in the large kill of waterfowl in the Great Plains. Toxaphene and endrin were implicated in the 30–40 percent decrease in endangered brown pelicans in Louisiana in 1974–1975. Each year more than a million cattle and sheep are treated with several million gallons of toxaphene solution to control parasites that cause mange or scabies. This is required by state and federal laws of sheep and cattle that have been exposed to mange-infected animals.

In 1977, the EPA began the long process of reviewing the chemical's suitability for continued registration. In 1982 the EPA restricted most uses of toxaphene, while allowing the livestock dipping program to continue.

Aldrin and Dieldrin

Aldrin and dieldrin were first used as pesticides in 1948, and because they were so effective, by 1950 they were heavily applied to such crops as corn, peanuts, potatoes, sugar beets, and cotton. These chlorinated hydrocarbons decompose very slowly in the soil, taking as much as 15 years to break down. There is strong evidence that these pesticides were overapplied to soils as simple preventive measures. By the time restrictions were put into operation, dieldrin was found to be present in 96 percent of all meat and poultry.

Aldrin and dieldrin are especially toxic. For example, it has been shown that a pill-sized portion of aldrin is sufficient to kill 400 quail. Both chemicals have been proven to be carcinogenic. By 1962, tests under way by the Food and Drug Administration showed that dieldrin produced cancer in laboratory animals at the lowest level at which it was tested, 100 parts per billion.

Although the Environmental Protection Agency has collected many years of records, hearings on the health effects of aldrin and dieldrin did not begin until August 1973. Immediately, the Shell Oil Company, the manufacturer of the chemicals, assembled a law team to protest the hearings. After a year of hearings, William Ruckelshaus, administrator of the EPA, suspended the manufacture of aldrin and dieldrin after August 2, 1974. The decision was based on the grounds that these chemicals posed an unacceptably high risk of cancer. The debate continued in the courts, but it finally ended in October 1974, when the presiding administrative law judge, Hubert L. Perlman, concluded that aldrin and dieldrin represented an unreasonable risk to the American public. Although most uses of aldrin and dieldrin have been banned, they are still permitted for termite control, dipping of nonfood roots and tops, and in mothproofing. And while the chemicals are no longer allowed for most uses in the United States, they are still used in some Third World nations. A less toxic chemical, chlorpyrifos, has been substituted for aldrin and dieldrin, with effective results, illustrating that many of the more toxic chemicals have substitutes that are equally effective.

Control Strategies

There are a number of ways to limit the amount of toxic chemicals in the environment. Some of these control strategies are discussed below.

Chemical Substitution

This is one of the most important control strategies, for it occurs at the initial stage. It has been proven that many less toxic chemicals are equally effective in controlling pests as are more toxic chemicals. A change in raw material input in the production process is required. For example, waterborne inks, while not a perfect substitute, can replace more toxic solutions such as toluene, acetone, alcohol, and others.

Product Reformulation

There are a number of opportunities to use less toxic compounds in the production process. The gradual elimination of lead-based paints and their replacement with solvent-based paints is one example.

Process Modification

Many toxic chemicals can be made in less hazardous ways. Changes may occur at many stages in the production of the product. For example, after the Bhopal accident it was discovered that methyl isocyanate can be produced at the rate it is consumed, thus eliminating the need for storage of the deadly substance. As waste management costs increase due to more stringent regulations, the reduction of both accidental and intentional release of environmental residues is becoming a fundamental reason for process changes. Generally, process modifications occur in the context of a specific process. To illustrate, the chloralkali industry, which uses electrolysis to produce chlorine, hydrogen, and sodium hydroxide, originally used a mercury cell, resulting in a large concentration of mercury and a smaller concentration of chlorinated hydrocarbons as wastes. A modification of this process used a diaphragm cell instead of a mercury cell, which eliminated the wastes.

Equipment Changes

The reduction of toxic emissions at the industrial source can often occur with redesigning or substituting equipment. For example, the use of air spray guns may release toxic chemicals into both the interior and the exterior atmosphere.

Operational Changes

Changing operation procedures can also reduce industrial waste output and thus minimize hazardous materials in the environment. Since many toxic materials appear in chemical plant wastewaters—heavy metals, sulfides, cyanides—reduction in the volume of these wastewaters would significantly reduce environmental releases.

Efficient Utilization

Traditionally there has been little control of the amount of toxic materials deposited on fields to control pests. Normal practice was to use excessive amounts. It has been demonstrated that the amount of pesticides used can frequently be reduced with no loss of protection. Thus, the users of pesticides need to be educated about amounts to use, and there needs to be greater controls placed on the amounts allowed in use.

Chemical Accidents

Major chemical accidents are rare, but when they do occur they create a disaster in the region. Control during the manufacturing process is thus essential. When major accidents occur, they receive worldwide attention.

Bhopal

On the night of December 2, 1984, and the early morning of December 3, water leaked into a tank at the Union Carbide Chemical plant in Bhopal, India. The tank contained methyl isocyanate (MIC), a highly toxic chemical compound. The intense reaction turned the liquid to gas, and in the ensuing explosion the gas flowed over the town of Bhopal, causing the greatest industrial disaster in history. The official figures place the number of deaths at 2,850 and the number of others seriously affected between 30,000 and 40,000. The pathological after-effects of the survivors are still conjectural.

The factory was established in 1969 to manufacture alpha-naphthol. This liquid was mixed with MIC to make the carbaryl insecticide known commercially as sevin. Methyl isocyanate is one of the most toxic chemicals known in its gaseous form. MIC and its degradation products have a high potential for damaging biological systems on both short- and long-term bases by producing heat energy through exothermic reactions.

The accident took place because of several factors. At the plant site, the MIC was stored in three steel 60-ton containers. On the night of December 2 the temperature rose in the tanks and the pressure increased beyond the danger point. The intense heat caused the concrete above the tanks to crack. The gas initially poured out at a rate of 180 pounds per minute and then increased to a maximum pressure of 720 pounds per minute. The safety measures—control valves and a water-spraying system—were faulty, inoperative, and/or inadequate, and the gas could not be neutralized or contained.

The major cause of immediate death was by asphyxiation, accompanied by severe irritation of the eyes and throat. During the first two days, the death rate in some of the worst affected areas was as high as 24 per 1,000 persons (compared with a record frequency of 1 per 1,000). At the time of the accident, medical teams were not able to determine the chemicals causing the toxicological effects. The nature of the chemical could not be specified, as to whether it was MIC, MIC and CO, or phosgene. No data were available to the doctors on the toxicology of MIC or on methods for dealing with exposure to MIC, and no data were available on the possibility of CO or CN being present through the dissolving of MIC. The initial absence of basic information on the nature of the leak and its possible degradation products delayed appropriate detoxification measures, thus raising the mortality level. The location and subsequent expansion of the Union Carbide factory in a densely populated region magnified immensely the dimensions of exposure to the escaping toxic gas.

To prevent any more disasters like that at Bhopal, the following recommendations are made. Most of these involve routine safety precautions, but in many areas of the world they are not implemented, for several reasons: because government, the company, or both are uninformed, careless, or callous, or because the workers lack adequate training.

1. Industries that produce toxic substances should have fully organized systems of neutralizing the chemicals around the clock, manned by trained personnel.
2. The levels of pollutants in the atmosphere and the soil and the rate of excretion should be measured routinely.
3. The capacity of the storage tanks for volatile substances should be inversely proportional to the toxicity of the chemicals. Each container should have safety valves and cutoff mechanisms.
4. All products should be tested for human toxicity.
5. To minimize the effects of accidental spills in transit, detoxifying measures should be documented in detail and the documentation made available to the general public.
6. The need for toxic chemicals should be fully assessed before permission is given for manufacture or storage.
7. The factory environment should be surrounded by a green belt, preferably of plants with known resistance to the products being manufactured.

Taluker, Geeta, and Archana Sharma, "The Bhopal Accident: Its After-Effects," in Shyamal K. Majumdar, E. Willard Miller, and Robert F. Schmalz, *Management of Hazardous Materials and Wastes* (Easton, PA: Pennsylvania Academy of Science, 1989), pp. 414–415.

Criteria for Evaluating Waste Sites

If the public health and environment are not to be endangered, hazardous wastes must be deposited only in appropriate places. It is most important that inappropriate sites be rejected for such waste disposal. For example, any area where the wastes can move quickly to an aquifer must be rejected. The following criteria were prepared for the New England Governor's Conference; the work was partially funded by a grant from the U.S. Environmental Protection Agency (Clark-McGlennon Associates, *Criteria for Evaluating Sites for Hazardous Waste*

Management: A Handbook on Siting Acceptable Hazardous Waste Facilities in New England. Washington, DC: Environmental Protection Agency, 1980):

Area	Criteria
Surface and groundwater	Contaminant flow period and water table
	Contaminant movement with groundwater
	Predictability of contaminant movement
	Potential impact on surface water
	Potential impact on aquifers
	Potential impact on public water supplies
	Possibility of site flooding
	Potential human exposure to treated wastewater
Air	Nature and predictability of pollutant movement
	Potential human exposure to air pollutants
Transportation system	Safety of transportation routes
	Distance between sensitive sites and transportation routes
	Potential for noise impact
Land use	Potential for impact on environmentally significant lands
	Proximity to residential areas or sensitive sites
	Compatibility with existing land uses and land use plans
Economic	Potential effect on property values
	Impact on existing or future economic activity
Other	Potential for earthquake activity

In the selection of a waste disposal site there are also human factors that must be considered. When a new site is chosen, the inevitable response of nearby residents and business owners is, "Not in my backyard" (or "NIMBY"). This response is predictable, for there is fear of the potential health hazard a waste site brings. In many instances new hazardous waste sites have been poorly chosen, and there are valid reasons for the rejection of the site. In other instances, fears that the site will cause a deteriorating environment and a health hazard are powerful reasons to reject the site.

By the late 1980s many of the waste sites in the nation were filled or nearing capacity. The major question of where to dispose of wastes has reached critical dimensions. In order to solve this problem the public must become better informed and involved in the hazardous waste management decision-making process. It has been suggested that this be accomplished by establishing a means of communicating factual information through a series of workshops and seminars. Questions, however, persist, for experience suggests that a majority of the public will be interested in hazardous waste issues only when a real or perceived threat exists, such as when a hazardous waste depository is sited in their immediate neighborhoods.

Unfortunately, when the issue of siting a hazardous waste depository arises, the public's interest is focused on protection, and the waste management needs of a city or state are often lost in the discussion. In addition, the efforts to educate the public are usually sporadic and are often too general to affect a broad audience. Equally important, prepared materials on the subject are often distributed only to a small group of individuals who are interested in the issue. The general public, therefore, continues to derive its information from the public news media, whose reports are frequently written by individuals with limited backgrounds and training to understand the problems. The management of hazardous wastes will be an issue as long as an industrial society exists.

Nevertheless, since hazardous wastes will continue to be produced, and new and improved waste management policies will be required, the siting issue will eventually have to be addressed. To provide opportunities for active public participation there is a need for a focused and long-term commitment to the development and full implementation of an effective public education program at the local, state, and national levels. The magnitude of the hazardous waste management issue warrants a commitment on the part of the nation's education community. The study of hazardous wastes should be included in the curriculum for both social studies and science classes in the nation's secondary schools. Future voters need to gain a perspective on the social, economic, and environmental factors involved in this important issue.

Will the existence of an educated public end the emotional reaction of "Not in my backyard"? Probably not. Whenever an individual or a community feels personally threatened, NIMBY will appear. Nevertheless, a well-informed public would be able to assess siting proposals from a significantly different viewpoint and would be able to evaluate the situation somewhat objectively. If decisions cannot be made by an informed public, the problem of the growing number of hazardous waste sites can only get worse.

Transportation of Hazardous and Toxic Waste

The transportation of hazardous material occurs much the same as normal movement of goods in the United States. These waste materials are produced and transported in every part of the nation. Well over 98 percent of hazardous material is transported without incident, but the potential for catastrophic consequences is great when an accident occurs. This potential has caused the public to be very concerned about regulating and controlling hazardous material transportation.

TABLE 1.8
Department of Transportation Hazard Classes

CLASS	EXAMPLES
Flammable liquid	Gasoline, alcohol
Combustible liquid	Fuel oil
Flammable solid	Nitrocellulose (film), phosphorous
Oxidizer	Hydrogen peroxide, chromic acid
Organic peroxide	Urea peroxide
Corrosive	Bromine, hydrochloric acid
Flammable gas	Hydrogen, liquified petroleum gas
Nonflammable gas	Chlorine, anhydrous ammonia
Irritating material	Tear gas, monochloroacetone
Poison A	Hydrorganic acid, phosgene
Poison B	Cyanide, disinfectants
Etrologic agents	Polio virus, salmonella
Radioactive material	Uranium hexafluoride
Explosives	
Class A	Jet thrust unit
Class B	Torpedo
Class C	Signal flare, fireworks
Blasting agent	Blasting cap
Other Regulated Materials (ORM)	
ORM A	Trichloroethylene chloroform
ORM B	Calcium oxide, potassium fluoride
ORM C	Cotton, inflatable life rafts
ORM D	Small arms ammunition
ORM E	Ketone polychlorinated biphenyls

Source: U.S. Congress, Office of Technology Assessment, *Transportation of Hazardous Materials* (Washington, DC: Government Printing Office, 1986).

Classification and Flow

The U.S. Department of Transportation (DOT) has prepared a classification of hazardous materials that are transported. Table 1.8 lists the classes and examples of the thousands of hazardous materials transported daily in the nation. Of these materials, gasoline is the simple commodity involved in the largest number of accidents. The U.S. Department of Transportation indicates that the most hazardous materials are poisons, explosives, and radioactive materials. These commodities are given the highest priority of care in transporting. Flammable materials account for the largest quantities moved.

The Office of Technology Assessment has estimated that about 1,649,000 tons of hazardous materials are transported each year in the United States. Of this amount about 60 percent moves by truck, 35 percent by water, and 5 percent by rail. This movement entails about 783 billion ton-miles of material (Table 1.9).

The mode of transportation used to carry hazardous materials varies with the commodity. Petroleum products, including gasoline, make up about half of all truck hazardous transport. The largest hazard classes transported by rail are flammable materials (26 percent), corrosive materials (25 percent), and flammable gases (12 percent). Hazardous commodity shipments constitute 55 percent of all domestic waterborne tonnage, and 85 percent of the tonnage is crude or pro-

TABLE 1.9
Estimated Transportation of Hazardous Materials by Mode (1982)

Mode	Vehicles/Vessels Used in Transport	Quantity of Hazardous Materials Transported Tons (millions)	%	Ton Miles (billions)	%
Truck	370,000 dry freight or flatbed 130,000 cargo tanks	9.27	59.83	93.60	11.95
Rail	115,000 tank cars	73.00	4.71	53.00	6.76
Water	4,909 tank barges	549.00	35.44	636.50	81.23
Air	3,772 aircraft	0.29	0.02	0.46	0.06
	Total	1,649.29	100.00	783.56	100.00

Source: U.S. Congress, Office of Technology Assessment, *Transportation of Hazardous Materials* (Washington, DC: Government Printing Office, 1986).

cessed petroleum. Chemicals account for 80 percent of the hazardous material shipped by air.

Regulation and Safety

The transport of hazardous materials is regulated by the Hazardous Material Transportation Act as administered by the U.S. Department of Transportation. This act empowers the secretary of transportation to designate those materials deemed hazardous and to develop regulations prescribing how such materials must be moved. The regulations are detailed, specifying the materials subject to regulation, type of packaging, labeling, marking, placarding, handling, loading, and route procedures to be used. In addition, instructions are given as to the mode of transportation.

A number of concerns have been given particular attention. Proper identification of the hazardous material is vitally important, so that if there is an accident, proper steps can be taken immediately. Packaging is another major concern of hazardous material transport regulations, because the type of packaging will prevent, to a large degree, the release of toxicants in case of an accident. The basic requirement is that packaging must prevent release of materials during travel, including minor accidents. The regulations provide detailed specifications for the design, construction, and testing of the package. The most stringent packaging regulations are those for radioactive wastes.

The greatest risk in the transportation of hazardous material is an accident. The U.S. Department of Transportation defines a transportation accident as an unintentional hazardous material release during loading or unloading, transportation, or temporary storage during transportation. According to the law, all accidental releases, except from bulk water transport, intrastate trucking, and specified minor spills, must be reported to DOT by the carrier within 15 days of the accident. These accident reports are then entered into the Hazardous Materials Information System (HMIS) database. There is some evidence, however, that not all accidents are reported.

In the period 1976–1984, the Hazardous Materials Information System reported 78,082 incidents, or an average of 8,676 annually. In the late 1980s the number of accidents decreased to about 5,000 per year. Highway accidents accounted for 86 percent of the total; rail transport accidents made up about 12 percent.

The overwhelming number of accidents on highways make it clear that there is a need to increase safety in this area. A number of different types of safety programs have the potential to reduce road accidents. The first of these is driver training, licensing, and certification. The single greatest cause of accidents has been shown to be driver error, which can usually be related to lack of adequate driver training, poor driving habits, and/or drug and alcohol abuse while operating a vehicle. One way to address this problem is to require a special operator's license for anyone driving trucks containing hazardous materials.

A second area often targeted for safety improvements involves inspection, maintenance, and retrofitting programs for hazardous materials vehicles. There is a need for a federal inspection system and maintenance program.

A third category of safety improvement is operation improvement. This involves such changes as the establishment and enforcement of safe speed limits on freeway systems, and especially on approach and exit ramps.

A fourth major safety improvement area is that of emergency response procedures. If a community is to handle an accident properly when hazardous material is released, it must take part in preaccident planning. The first step in such planning is for community leaders to know the locations of all businesses and industries in the area that use, generate, or manufacture hazardous materials. A second step involves special training for individuals who will be expected to respond to any incidents. There is also a need for detailed traffic rerouting plans, as well as an emergency vehicle access system. The final area for preplanning is the development of further response capabilities. With effort and planning, a community can develop a hazardous materials accident response team. Such teams need to be highly trained, skilled units; they are usually associated with the community's fire department.

Emergency Preparedness

Although the overwhelming proportion of hazardous materials shipments reach their destinations without mishap, the need to handle emergencies does arise. Accidents occur most frequently at the terminals where the shipments originate and terminate, but may occur anywhere along the route. Clearly, rapid emergency response to spills of toxic materials is essential. Because of the potential danger from a toxic spill accident, a bewildering number of regulations have arisen at the federal, state, and local levels. At each of these levels, groups are engaged in planning for emergencies, providing emergency response services, and training personnel who are to respond to the emergencies.

While planning for emergencies has developed at all government levels, the single common feature of the majority of accidental releases of hazardous materials is that the local community must provide an immediate response to a particular incident. Thus the local community and its agencies have the primary responsibility for handling these incidents. This does not minimize the responsibility of federal and state agencies, but rather emphasizes the importance of local preparedness and response.

The local government and its agencies occupy a portion within the overall response that cannot be filled by any other organization. They have a vested interest in the progression and outcome of any incident. Because they are familiar with local conditions, they have a responsibility to provide for and protect the best interests of the community. Local officials must evaluate, coordinate, and prepare in order to fulfill the public trust that they have been granted.

In the development of emergency planning the first commitment of the local official is to develop a preplan. The process of developing a risk analysis is very basic. This initial step involves finding out the exact situation of the community with respect to the types, quantities, locations, and hazards of materials that can be found in the community at any time. This also involves routes for the movement of hazardous materials through the community.

As a next step, it is important to establish a management organization in which individuals or groups have specific responsibilities. These people will make up the nucleus of the Emergency Operations Center. This group usually includes elected officials, fire coordinators, the police chief, a public information officer, the public works coordinator, a community officer, and the community engineering department. All of these agencies must be coordinated into a working team. In most localities, the police have a system developed. It is important to review existing systems to ascertain their practicality and usefulness in a hazardous material incident.

In order for emergency planning to function effectively when a hazardous material incident occurs, the individuals in the management system require commitment and training. Tests must be conducted on all phases of the operations.

The next step is to prepare a master resources list, including human resources, materials, equipment, services, and organizations that are available to the community and that will aid in handling all aspects of an emergency. This list must be precise, readily accessible, and prepared in advance of need. In the process of establishing the resources list, it is important to ask several questions concerning what types of resources are available through local government agencies, police organizations, fire departments, and so on.

Experience has shown that a hazardous materials emergency response team composed of well-trained individuals is the best way to handle a hazardous materials incident. In order to assure that the emergency plan will function during a real incident, the team should run tests using simulated situations. Such drills can help the team learn how best to evaluate an accident situation.

Finally, critiques are a major factor in developing, maintaining, and evaluating the effectiveness of the plan as well as the overall operation of the agencies involved. Critiques should become part of the standard operations of the incident management system. Without the use of critiques, a major portion of its learning and evaluation process will be missed. These critiques can also point out shortcomings and improper procedures that can be corrected.

Emergency response plans to deal with hazardous materials incidents have now been completed for thousands of communities. Such planning was made a national priority in the Superfund Amendments and Reauthorization Act of 1986. Title III of SARA requires each state to establish an emergency response commission to oversee the development of local emergency plans. The *Hazardous Materials Emergency Planning Guide,* prepared under the auspices of Title III, presents an excellent summary of the act's planning requirements and gives detailed guidance for implementing these provisions. The Office of Technology Assessment estimates that there are 2 million people in the emergency response network, but only 25 percent have received adequate training.

Effects of Hazardous Waste

There are a number of potential effects of hazardous waste sites on the areas near them. Four basic kinds of effects are health, social, economic, and environmental; each of these is discussed in turn below.

Health Effects

The effect of toxic substances on health has received much attention from the research community. There have been many studies about health problems resulting from exposure to toxic substances, but despite the importance of being able to predict outcomes of the disposing of hazardous chemical wastes, uncertainty about data, methods, and effects remains. It is difficult, if not impossible, to correlate abnormal numbers of cancers, adverse reproductive cases, or other acute chronic diseases with hazardous waste sites.

There are a number of reasons this uncertainty persists. A major factor is that the hazardous characteristics of most chemicals have not been tested. Relatively few chemicals have been tested for toxicity, and still fewer for mutagenesis, teratogenesis, and carcinogenesis. It has been argued that since only a few chemicals dominate the market, testing should be limited to these few. This reasoning is fallacious, for in most hazardous waste sites a very wide variety of chemicals are found.

It is not feasible to test all chemicals for epidemiological or annual effects. Only the application of short-term tests is realistic for many chemicals in the near future, and it is difficult to extrapolate short-term tests to long-term human reactions. There is also a problem in determining the effects of chronic, low-dose exposures. In the absence of data, scientists debate the relative validity of linear dose-response models that produce considerably more health effects than do models that assume that there is a lower threshold before an effect will occur.

Some chemical substances have been tested, but scientists are reluctant to make predictions because of possible limited or incorrect results. For example, it was once thought that tricholomethanes in drinking water increased bladder cancer. After more detailed study the National Academy of Sciences concluded that it was impossible to establish such a linkage.

Another major cause of uncertainty about the impact of hazardous waste disposal sites on public health is variation in human receptivity. Studies have shown that there are toxins in many humans, but they produce wide variations in health effects. A high body burden of toxins does not lead to health problems in all persons. Some humans are extremely resistant to toxic substances, and others are not.

To summarize, research on the health effects of the toxicity of waste sites is still at an early stage of development. There can be no doubt that the effects of various levels of toxicity can range from minor illnesses to cancer or death. But before we can draw meaningful conclusions, there needs to be a great deal more known about the hazardous characteristics of substances at chemical waste sites, the diversity of chemicals found there, and how the human population reacts to exposure to those chemicals.

Social Effects

There are numerous stories of the social effects on people living near toxic waste sites. In a few instances whole towns have been evacuated. In the process, family units may be broken and the social infrastructure destroyed. In essentially every survey thus far conducted, Americans have said that they do not want to have hazardous waste sites in their neighborhoods. Solid waste management is a relatively recent issue, having arisen as city landfills are filled and sites for waste disposal are sought in distant areas. A city with a hazardous waste facility is now perceived as an undesirable place to live.

Economic Effects

The impact of solid waste sites on economic values has received limited attention. It is normally assumed that the impact is negative. Studies have shown that persons with homes having higher property values were more likely to be opposed to having a waste site nearby than persons with homes of lower property value.

One economic effect of living near a hazardous waste disposal site can be increased taxes. When the Environmental Protection Agency recommends the removal of buildings and cleanup of land at waste sites, it always attempts to make the owners of the property liable for the expenses, but this is not always possible; sometimes the company has disappeared. Because the cleanup of hazardous sites is very expensive, when cities must absorb the cost, tax revenues must be increased.

Environmental Effects

When there is toxic contamination of soil and water, the total ecosystem of an area is in danger. Reduction or elimination of certain plant and animal species may occur. There is also the possibility of a reduction in the diversity of life. Further, the loss of soil reduces the productivity of the area, and the contamination by waste resources makes it unusable for animals or humans.

Summary

It is difficult to draw accurate conclusions about the effects on health, economics, social conditions, or environmental conditions of hazardous waste from study of a few special sites. Many of the earlier writings on this topic based conclusions on very flimsy, unscientific evidence. The lack of reliable health data does not provide a sound foundation for drawing conclusions about potential health effects. Social and economic effects are presumed to be negative, but have not been systematically studied. The public can arrive at wrong conclusions because of lack of basic information or, even more important, emotionally biased statements.

Petroleum Environmental Pollution

Petroleum, when spilled on land or in rivers or the oceans, creates an environmental hazard. Oil pollution has not occurred in amounts significant enough to affect the environment until recent decades. Only since oil has been produced in large quantities has the environment been threatened. The first great oil spills occurred at sea, but because oceanic pollution affected few people directly, the dangers were long ignored. Even as late as the 1950s, no international law protected the oceans. Only when pollution became a "social cost" did it become a political issue. The first international controls were initiated in 1954, and since then a vast body of regulations has evolved. Most of these regulations have been initiated by the Intergovernmental Maritime Consultative Organization (IMCO). There is now a body of rules and regulations on accidental pollution and a detailed intergovernmental scheme for assigning liability when prevention fails. While progress has been made on international oil pollution, the problem of pollution due to accidents remains a critical issue, as demonstrated by the 1989 spill of some 240,000 barrels of oil due to accidental grounding of an oil tanker at Valdez, Alaska.

Sources of Oil Pollution

Oil pollution can occur on land or in the oceans. There are three major sources of oceanic oil pollution: natural seepage from underground deposits, spills from tankers caused by operational activities and accidents, and oil carried to the ocean by stream runoff. Land sources of oil pollution are accidental spills from tanks and pipelines and improper disposal of waste oils. There is also danger of oil spills from oil exploitation, on either land or water.

Natural Seepage from Underground Deposits

In many places in the world, oil is able to penetrate the surface through fissures from oil sands. Because many oil fields lie beneath oceanic waters, this oil escapes into the water. This is a natural phenomenon and will continue indefinitely. There is no reliable way to know exactly how much oil escapes this way into the land and into the oceans annually, but estimates range from 1,500,000 to 4,000,000 barrels.

Oceanic Pollution from Land Sources

The major source of oceanic oil pollution in the past was oil pollution on land. Much of the oil spilled on land is ultimately carried to the sea by streams. In the past there was waste oil from oil refineries and petrochemical plants. These oils were frequently discharged accidentally or deliberately into the river systems. Since the passage of the Clean Stream Act of 1970, the deliberate pollution of streams has essentially stopped, and accidental pollution has been reduced.

Spills from Oil Exploitation

Oil spills may occur during drilling of oil fields, either on land or in oceanic settings. When this occurs in oceanic waters, the oil is distributed over a much wider area and the environmental damage may be great. In the early days of drilling it was not uncommon for the oil pressure in the producing formation to be greater than expected and for spills to occur as the oil-well casing fractured. Little attention was paid to these events until an oil spill that occurred in the waters off Santa Barbara, California, on January 28, 1969. In 1977, another major oil spill occurred at Ekofisk in a North Sea drilling operation. In each of these spills about 700,000 barrels of oil were spilled. Public awareness of these accidents was great, and demands to protect the environment were spurred.

Recent data indicate that in the offshore waters of the United States, 1 well in 1,000 will spill small amounts of oil, but only 1 well in 10,000 will spill sufficient quantities of oil to create an environmental hazard. Even under the best control conditions, some oil will thus be spilled. It is now estimated that spills from the operation of the world's undersea oil fields average about a million barrels annually.

Operational Pollution from Tankers

Most of the oil fields of the world are located far from the major oil-consuming countries. The oil must thus be transported, as either crude or refined products. For most of these oil fields, such as those of the Middle East, North Africa, Nigeria, and Venezuela, the only means of shipment is by tanker. As a result of the growing demand for petroleum, the world movement of oil increased from a little less than 3 billion barrels in 1960 to about 12 billion barrels in the late 1980s. Although oil spills may occur in the operation of the oil tankers, such as in tank cleaning, ballasting, and oil transfer from tanker to land, the greatest danger of oil spills comes from tanker accidents.

Tanker accidents can result from a number of causes. Structural failures of vessels account for about one-third of the accidents. To illustrate, on March 16, 1978, the *Amoco Cadiz,* carrying 1.5 million barrels of light Arabian crude, experienced rudder failure when approaching the English Channel. Before a tug could reach the vessel, the weather deteriorated and the tanker grounded on the Brittany coast. On March 24, the vessel split in two, spilling the entire cargo. A large proportion of the oil was carried onto the coast of Brittany.

The second most frequent cause of oil tanker accidents is the grounding of vessels. The first supertanker to ground was the *Torrey Canyon* on the Seven Stones Shoal about 20 miles off Cornwall's Land's End on March 18, 1967. The tanker spilled 830,000 barrels of Kuwaiti oil into the sea. This was the world's first major supertanker accident; since then, there have been groundings of many large tankers. The most recent cargo spill occurred when the *Exxon Valdez* grounded on Bligh Reef in Prince William Sound on March 24, 1989, spilling 240,000 barrels of crude oil in the harbor. This grounding was due to human error. The captain of the vessel was not in command as required by the Coast Guard license; the vessel was in the hands of a relatively inexperienced third mate. Because radar indicated icebergs in the area, the third mate requested permission to change his route, setting the disaster in motion.

The third major cause of accidental oil spills at sea is the collision of tankers. A spectacular collision between two tankers, the *Ven Oil* and *Ven Pet*, occurred December 16, 1977, off Port Elizabeth, South Africa. One vessel was loaded, but the other was in ballast. As a result of the collision some 210,000 barrels of oil were spilled, polluting a considerable area of the South African coast. Collisions are usually due to negligence. If the sophisticated navigation equipment carried by tankers, particularly the supertankers, is used properly, collisions are avoidable.

Explosion and fire constitute the fourth major hazard causing tanker spills. These frequently occur after another accident has damaged a vessel. For example, the *Jacob Maersh* struck an underwater obstruction while trying to berth at Oporto, Portugal. This accident triggered an explosion and fire, and the entire cargo of about 560,000 barrels of crude oil was either burned or spilled into the sea.

Land Oil Spills

Petroleum crude oil and refined products are stored in bulk storage facilities. When a tank fails, an oil spill of major proportions can occur. One of the major inland spills to date occurred on January 2, 1988, at Floreffe, Pennsylvania, on the Monongahela River, when a tank holding 4 million gallons of no. 2 fuel oil ruptured. The oil flowed through the storm sewers into the river and mixed with the water as it tumbled over the dams, creating an emulsion. This emulsion did not stay on the surface, but contaminated the entire river, from the surface to the bottom. Because the towns along the Monongahela and Ohio rivers obtained their water supply from the river, the water intake valves were closed from the site of the spill to the Point in Pittsburgh and along the Ohio River to the border of the state. The governor of Pennsylvania declared a disaster emergency, and the Emergency Operation Center was fully activated, with all pertinent agencies present. By the end of the first day of the spill more than 13,000 people were without water service to their homes. Water was supplied by truck to the affected communities. The lack of water raised special concerns in the areas of fire fighting, hospital needs, homes heated by hot water and boilers, and electrical power facilities. By January 11, 1988, most of the oil had been carried downstream, and through evaporation and mixing with the river's water, the danger of contamination ended.

At the conclusion of this incident a process was established to identify procedures for handling such an accident. Because major oil spills are relatively rare, management techniques and controls may become lax. If this occurs, a real tragedy is in the making.

Disposal of Waste Oils

In our mechanized modern society millions of barrels of oil are used each year in lubricants, antifreeze solutions, transmission and brake fluids, and so on. A significant proportion of these oils is not consumed or destroyed, but remains as waste residue. In 1974, a report by the Environmental Protection Agency estimated that more than 28 million barrels of waste oil are generated annually in this country.

These waste oils are disposed of in a number of ways. An estimated 43 percent are burned, usually with little regard to air pollution. An additional 31 percent of the oils cannot be accounted for. A large proportion of waste oil originates in tiny amounts at thousands of points in the nation, at such places as garages, service stations, industrial plants, and individual households. Most of this oil is dumped on the ground, where part of it seeps into the soil and the remainder is washed into the streams and ultimately deposited in the oceans. This vast problem can be solved only by educating the public about the dangers of waste oil to humans and the environment.

Waste oil has traditionally been used to reduce the dust from roads and open spaces. It is estimated that 18 percent is spread on land surfaces. There are hundreds of thousands of miles of unpaved roads in the United States. During summer weather particularly, dust raised by auto traffic on these roads creates a major nuisance. A report of the Congressional Research Service of the Library of Congress indicates that at least 120 million gallons of waste oil are spread over rural dirt roads annually in an attempt to control this dust.

Waste oil that is dumped on the ground or spread on roads can pose a significant threat to human health and the environment. Spent motor oil, for example, contains high concentrations of lubricant additives such as detergents and heavy metals—lead, zinc, phosphorous, barium, and vanadium—and traces of metal from the motor. These oils also contain significant concentrations of carcinogenic polycyclic aromatic hydrocarbons (PAH), PCBs, and concentrated pesticide wastes.

Only about 8 percent of waste oil is currently cleaned and re-refined for commercial consumption. Although there are between 1,000 and 2,000 waste oil collectors in the nation, they are located almost exclusively in urban areas, and most are poorly equipped. The economically efficient collection of waste oil is difficult because it is so widely dispersed. No one place has a sufficient quantity to justify the building of a collecting system and plant operation.

As early as 1972, Congress became concerned about the disposal of waste oil and requested the EPA to study the problem. Although the EPA has the authority to control waste oil under the Resource Conservation and Recovery Act, the agency believes that any attempt to enforce controls against small producers such as service stations would be ineffective; its regulations are directed at large producers. This type of regulation is weak and largely ineffective.

Little progress has been made in controlling waste oil disposition. In 1980 Congress passed the Used Oil Recycling Act, but because of small appropriations to implement the act, limited development has occurred. There is a fundamental need for statewide collecting systems, coupled with strong enforcement against unsafe dumping. Waste oil should not be dumped, but should be cleaned of unwanted contaminants and then re-refined for commercial use. Huge quantities of waste oil could be recycled in this manner on a regional basis.

Control of Oceanic Oil Spills

Whenever an oil spill occurs near a coast, the oil endangers the shoreline. A number of techniques have evolved to combat the effects of such spills.

Attempts to disperse oil at sea before it reaches the coastal area have not been successful. As oil spewed from the wreck of the *Torrey Canyon*, the Royal Navy used about 700,000 gallons of detergents to break up the oil while it was at sea. At Santa Barbara chemical dispersants were also used to break up the oil. The conclusion reached was that these types of chemicals are ineffective in controlling large amounts of spillage in the open sea.

Mechanical devices to control the spread of oil in the open sea have also been ineffective. These included booms, skimming devices, and suction pumps. Booms have proved wholly ineffective in open water, because they break up in even modest seas and are difficult to maneuver. In harbors booms are more successful in reducing the impact of the oil on shore by repelling the oil before it can enter the protected areas.

Because accidents are not expected, a major problem has been the assembly of equipment. For example, during the *Exxon Valdez* spill the only contaminant barge in Valdez was in dry dock for repairs. Oil skimmers capable of recovering 50 percent of the spilled oil were to be at the scene of the accident within 5 hours, but because of delays they were not able to operate until 18 hours after the spill. In addition, company officials debated for two more days, and experimented with ways to deploy chemical dispersants and cope with potential fire. By the time recovery efforts began, the oil slick extended across nearly 40 miles of the 70-mile-long sound. Oil was spreading onto the beaches by that time.

Attempts have also been made to find materials that will absorb the oil as it reaches the shore. Talc and perlite have been somewhat successful, but straw and sawdust have proved to be most useful. On rocky areas pressurized hot and cold water and steam cleaning have been somewhat successful. Sandblasting has proved to be the only really satisfactory method of cleaning oil-stained rocks.

When a coastal spill occurs, evidence of the oil remains on shore for many years. In time, the oil is dissipated into the sea by wave turbulence, blowing new sand, and gradual penetration into the beach sand.

Ecological Damage

The ecological damage of fauna and flora after an oceanic oil spill varies greatly. Although a large number of studies have been made on the effects of oil spills, there are conflicting results on long-term damage. A continuing effort to study the effects of oil pollution in the oceans and coastal areas over an extended time span is needed.

Birds

As oil reaches a shoreline, birds suffer the most evident immediate loss. In the Santa Barbara oil spill it was estimated that a minimum of 6,000 and possibly as many as 15,000 birds died as a result of oil contamination. In the *Exxon Valdez* tanker accident U.S. Fish and Wildlife Service biologists estimated that within one week after the spill more than

15,000 seabirds had been exposed to the oil. There was growing concern about the impending arrival of millions of migratory birds that flock to the area each spring.

In all coastal disasters there have been major attempts to save the birds. However, the survival rate is amazingly low, usually 1 to 11 percent. The seabirds die from shock, from improper cleaning and handling, and from ingestion of oil and toxic chemicals during preening. Many birds also die of starvation and exhaustion, because they are unable to hunt for food. In the *Torrey Canyon* spill the Plymouth, England, laboratory reports indicate that human-made pollutants, such as detergents, were far more damaging to the ecology than the oil from the spill.

Marine Life

Of the oil spills affecting the coastal United States, the Santa Barbara oil spill is best documented. The effects of that spill on marine life varied greatly. Barnacles and surf grass suffered significantly. Some colonies of acorn and goosenecked barnacles were virtually destroyed. In contrast, anemones, mussels, limpets, and starfish were little affected. Common amphipods and bloodworms dug into the sand and escaped injury. Bottom studies revealed no damage to sea-floor plant or animal life attributable to the oil spill. The plankton declined initially, but recovery was rapid.

There was inconclusive evidence on the damage to fish and shellfish. Statistical surveys conducted by state, federal, and academic investigators showed no significant damage to the fish population. The major problem in determining the spill's effect on the fish population was that there was little information available on the movements of migrating fish. Large numbers of anchovy, mackerel, and bonito entered and left the channel during the spill's early months.

Mammals

In the *Exxon Valdez* spill concern was focused on the area's 5,000 sea otters, which are especially vulnerable to oil because it reduces the insulation and buoyancy of their fur. If they do not die of cold, they can ingest fatal doses of oil while licking it off their coats. Of those caught and cleaned in the weeks following the accident, only a few lived. Hundreds of others died slow deaths on remote oil-fouled shores.

In the Santa Barbara oil spill, evidence of the effect on mammals was not conclusive. The Santa Barbara area is noted for its California sea lions. It appeared that the mature animals were little affected, but many more of the pups that were oily died than would have been expected. There was also evidence that several whales died from oil contamination. Whales and other sea mammals are highly susceptible to respiratory ailments, and any interference with their respiratory systems can result in illnesses such as pneumonia.

Development of Environmental Laws of the Ocean

In order to control oceanic pollution, rules and regulations must be based on understanding of the pollution problems, the economics and politics involved in each of the world's nations, and, finally, the effectiveness of the organization that implements the rules and regulations.

Defining the Problem

There are three aspects that must be considered in the development of control of international oil pollution. First, the importance of oil pollution in the oceanic environment must be determined through sound scientific analysis. Second, the oil pollution in the total global energy system must be assessed. Third, the technology required to control oil pollution must be developed.

Most studies on the impact of oil in the oceanic environment do not provide conclusive evidence of the harm done to the environment. In localized areas the immediate damage of a large oil spill is evident, but the environmental effect of gradual pollution from land sources has been largely ignored. As a consequence, there has been little urgency for politicians to develop stringent controls that are frequently costly. When conflicting scientific studies appear, legislators have had a tendency to support the least costly approach.

The second problem in developing regulations to control oil pollution is that it is sometimes considered a rather insignificant aspect of the total global energy system. The great growth in oil consumption due to economic development is the ultimate cause of the oil pollution problem. As a consequence, the means of controlling pollution are limited to what the larger economic system will permit. Because the system is international in scope, there is not only great competitiveness but also interdependence within the system.

Intertwined with the economic problem is that of technological development. Because of their economic and political power, the major oil companies exert a strong influence on the development of regulations. In general, industry has been reluctant to develop new control systems, except under great political pressure. For example, the double-hull vessel that would prevent most oil spills caused by accidents is available, but because of high construction costs, it is not widely used by the oil industry's tanker fleets.

Oil Pollution Regulation

The first international controls on shipping began in the World War II period, when the United Maritime Authority (UMA) was established. In 1948, the UMA was succeeded by the United Maritime Consultative Council (UMCC). In that year governments were invited to attend the U.N. Maritime Conference in Geneva to consider the establishment of an international body within the United Nations. Within 17 days the convention created the Intergovernmental Maritime Consultative Organization. However, it was not until 1958 that the required 21 nations ratified the convention and the IMCO began to function.

The IMCO does not have the power to enact laws. As the word *consultative* in its name implies, it can pass recommendations, convene conferences, draw up conventions, and facilitate consultations among member states. The resistance to the IMCO as a lawmaking body was based on the grounds that this action could introduce political distractions into an area that was fundamentally technical in nature. Although the initial convention of the IMCO did not mention a pollution control function, by 1965 pollution control aspects became so important that the organization established a subcommittee on oil pollution under the Maritime Safety Committee. The increased involvement in pollution control has given the organization viability. The IMCO has recommended a vast number of regulations aimed at the control of oil pollution of the oceans. Individual nations have made many of these regulations laws.

Research Needs

After the *Exxon Valdez* oil spill, it became apparent that determination of the ecological impact of crude oil on the environment was hampered by a lack of scientific information on birds, fish, and other wildlife in the area. Basic information needed to assess the impact on the ecosystem was found to be sorely lacking. A hodgepodge of scientific studies have been completed, but no one has developed a comprehensive analysis.

To correct this deficiency the U.S. Environmental Protection Agency in 1990 began a long-term project known as EMAP—the Environmental Monitoring and Assessment Program. The program will collect basic scientific data on the nation as a whole. When this information is placed on a grid it can provide basic environmental and ecological information for any ocean area around the United States. If, for example, a tanker were to run aground in the Philadelphia area, scientists could immediately secure information about the natural resources of that area. EMAP will become a crucial tool that will be vital for decision making in ecological disasters.

Asbestos Control

The use of asbestos fiber to control fire has had a long history, going back at least to Roman times. During the Industrial Revolution it was used for insulation. Asbestos deposits are widely distributed, but the province of Quebec in Canada has about 70 percent of the world's production of about 4.3–4.5 million tons annually. Demand for asbestos is now less than one-third of what it was at its peak in 1973 as a result of fear of toxicity and health-related problems. In 1971, asbestos became the first material to be regulated by the Occupational Safety and Health Administration (OSHA).

Definition and Transport

The word *asbestos* refers to the fibrous varieties of a family of naturally occurring silicate minerals such as chrysolite, amosite, erocidolite, tremolite, anthrophyllite, and actinolite. These minerals have the ability to be split into fibers that have such advantageous commercial properties as stiffness, high tensile strength, heat and chemical resistance, and electrical nonconductivity.

After asbestos is crushed, the particles are so tiny that they are easily transported by both atmospheric and water systems. Because asbestos fibers are fine and have little weight they can be carried great distances by air currents. Further, because of their light weight the gravitational pull on them is slight, and they remain suspended in the atmosphere for long periods. Asbestos may also be transported long distances in water systems. One study revealed that asbestos tailings from taconite deposits near Silver Bay, Minnesota, traveled more than 120 miles in the calm waters of Lake Superior before settling. The same particles traveled 294 miles from the source when vertical turbulence occurred. Thus people located hundreds of miles from asbestos sources may be subject to asbestos contamination.

Health Effects

The effects of asbestos on health have long been recognized. Ancient literature provides accounts of "sickness of the lungs" in slaves who worked with asbestos. As early as 1907, pneumoconiosis, later know as asbestosis, was associated with asbestos.

Asbestosis is a chronic, restrictive lung disease caused by the inhalation of asbestos fibers. It is characterized by the scarring of the lung tissues so that the pulmonary functions are impaired. The most prominent symptom is breathlessness. Asbestosis is associated with long-term high exposure to airborne asbestos. It is a progressive disease, with varying degrees of severity that increase with asbestos exposure. Even after exposure to asbestos stops, the effects can continue to increase due to the residue in the human system.

As early as 1935, a study of South Carolina textile workers indicated that asbestos could cause lung cancer. In 1947, another study found a high number of lung cancers in workers who had died of asbestosis. While lung cancer is caused by many conditions, it has been found that asbestos insulation workers who smoked cigarettes had an eightfold greater risk of bronchial carcinoma deaths than cigarette smokers who did not work with asbestos.

Development of lung cancer in asbestos workers may require more than 20 years, although the duration of exposure to asbestos fibers may be a very short period. Therefore, the relationship between the two is sometimes difficult to establish. There does seem to be a relationship in that the longer the exposure to asbestos, the greater the risk. Most epidemiological studies have indicated that there is no threshold for development of lung cancer from asbestos exposure. It is thought that a fiber's size (length versus diameter) is more important than the type of fiber. Fibers must be of a certain size in order to be respirable and ultimately deposited in the lungs and therefore capable of causing lung cancer.

The Environmental Protection Agency currently estimates that there are between 3,300 and 12,000 deaths each year from cancer caused by asbestos exposure. A program introduced in 1966 to control the use of asbestos will prevent only some of the expected deaths in the future for several reasons. First, some asbestos products will continue to be used. Second, and most important, millions of people will continue to be exposed to asbestos existing in buildings. Finally, it can take decades for people exposed to asbestos to become ill.

The most serious cancer caused by asbestos is mesothelioma. This is a rare form of cancer that appears as a thick, diffuse mass inside the serous membranes that line body cavities. Several epidemiological studies have indicated that mesothelioma is located at two sites: the pleura, which is the serous membrane that surrounds the lungs and lines the thorax, and the peritoneum, the serous membrane that surrounds the abdominal organs and lines the abdominal and pelvic cavities. Between 85 and 90 percent of mesotheliomas are associated with asbestos, but a few rare cases have been found in people who had no asbestos exposure.

Asbestos is known to cause other cancers as well. Studies have shown that with the inhalation of asbestos cancer has developed in the gastrointestinal tract, larynx, oral cavity, pharynx, pancreas, kidneys, and ovaries. It has been proved that asbestos fibers can reach any anatomic site in the body, increasing the risk of cancer.

Regulation of Asbestos

Six federal agencies are responsible for the regulation of asbestos.

1. The Occupational Safety and Health Administration is responsible for establishing limits for exposure in the occupational environment in order to ensure the health of the worker.
2. The Food and Drug Administration prevents asbestos contamination in foods, drugs, and cosmetics.
3. The Consumer Product Safety Commission regulates asbestos in consumer products. Asbestos is now banned in drywall patching compounds, ceramic logs, and clothing.
4. The Department of Transportation regulates the packaging and shipment of asbestos via public highways, rails, and vessels.
5. The Mine Enforcement and Safety Administration sets standards for mining of asbestos.
6. The Environmental Protection Agency has banned almost all uses of sprayed asbestos materials and asbestos that is friable. Studies are under way to consider banning all non-essential uses of asbestos for nonconsumer products. The EPA has charge of the asbestos abatement program in schools. In 1982 the agency issued the Asbestos-in-Schools Identification and Notification Rules, which require school officials to inspect for friable asbestos materials, to notify parents and teachers if such material exists, to place warning signs in schools where asbestos is found, and to keep records of their actions to assess conditions. To assist schools, Congress passed the Asbestos School Hazard Abatement Act of 1984. This act provides both technical and financial assistance to local educational agencies.

Asbestos Control

Because asbestos was used in the past to insulate a wide variety of buildings, and particularly school buildings, its control received wide attention in the 1980s. No other toxic material has had so much specific federal legislation enacted to control its use. In addition, the legislation not only controls future use but establishes guidelines for removal of existing asbestos. There are three accepted procedures to use in removing asbestos from existing sites—removal, encapsulation, and enclosure.

Removal

In the removal process the asbestos is stripped from the underlying surface, collected, and placed in containers for burial in an approved disposal site. The removal may use a wet or dry process. In wet removal the asbestos is treated with a water solution containing a wetting agent to increase fiber release. A wet method gaining popularity is hydroblasting of the asbestos surface with a low-volume, high-pressure water delivery system. Some asbestos materials are resistant to water applications, and for these a dry process is used. In this procedure the asbestos is scraped off with chisels and wire brushes. This technique is effective, but it is also difficult and time-consuming. Many experts believe that removal is the only effective way to control asbestos in the long run.

Encapsulation

In this technique the friable asbestos material is coated with a sealant. This is the quickest and least expensive method of controlling the fibers. In the process, the sealant may penetrate and harden the asbestos material, or a bridging encapsulant may cover the surface of the material using airless equipment at low pressure in order to reduce fiber release during the application. Encapsulation is frequently used when the asbestos is not readily accessible. It is also used as a precaution against future deterioration on dense, hard material containing asbestos that is not subject to water or impact damage. Encapsulation is not possible if the asbestos-containing material has poor internal cohesive strength, is friable, is damaged by water, or is firmly attached to the underlying surface. Encapsulation is not a final solution; it only delays the problem of disposing of the asbestos to some future time.

Enclosure

In this technique, a barrier is placed between the friable asbestos and the building environment. For example, a suspended ceiling may be built beneath an original asbestos ceiling, or metal casings installed around exposed asbestos pipes. This is a rarely used option, because the asbestos remains in place and has the potential for exposure during routine maintenance.

Legal Action

When asbestos was declared a carcinogen, more than 130,000 cases were filed in U.S. courts against hundreds of former asbestos manufacturers, distributors, and users. Of these companies, the Manville Corporation was for decades the world's largest asbestos manufacturer and became the largest defendant. This translated into litigation, and by 1982, this single corporation was faced with more than 16,000 lawsuits and increasingly large payments as the result of unfavorable jury verdicts. In August 1982 Manville sought the protection of the bankruptcy court.

After long litigation in the federal bankruptcy court in Manhattan, the Manville Corporation on December 9, 1988, established the Manville Personnel Injury Settlement Trust to provide payment to victims of asbestos exposure. Under this bankruptcy plan, Manville believed it would be shielded forever from all legal liability in connection with asbestos, in exchange for periodic contributions to the trust. By July 1990, Manville had paid $974 million to 22,386 claimants, or an average of $43,509 a claim. At this rate the trust would have to pay out more than $5 billion, although the total assets over its lifetime are not expected to exceed $3 billion. The trust had depleted all of its funds, and new funds could come only in the future.

Legal scholars are calling the handling of the trust funds the largest product-liability disaster in American history. It is now contended that the trust is a casualty of bad management, poor planning, and lack of oversight by the bankruptcy judge. In an agreement of the committee of the plaintiffs' lawyers, there was to be no scrutiny of their compensation. Personal-injury lawyers typically receive a third or more of individual settlement fees or court awards. The initial intention was that fees were to be much lower because the volume of cases was to be much higher. In fact, the plaintiffs' lawyers continued to charge high fees, 33–40 percent, all without question from the trustees of the fund. One lawyer presented a bill to Manville for $2.3 million.

As a result of the depletion of the trust fund, tens of thousands of asbestos victims will receive little or no compensation in the immediate future. The Manville Corporation believes that it bears no responsibility for bailing out the trust beyond operating according to the terms of the original bankruptcy plan. The bankruptcy proceeding calls for (1) payment of $1.8 billion for amount of bonds issued by Manville or $75 million per year for a maximum of 24 years beginning in 1991;

(2) starting in 1992, up to 20 percent of any profit Manville makes (the company is now highly profitable, producing fiberglass and wood products); (3) 7.2 million shares of Manville series A preferred stock convertible into 72 million shares of common stock (no stock can be sold within five years); and (4) 24 million shares of Manville common stock (50 percent of the company's outstanding stock in 1988 when it emerged from bankruptcy).

In order to solve these problems, federal Judge Jack B. Weinstein—who in July 1990 heard the case of 500 plaintiffs who say they suffer from inhaling asbestos at the Brooklyn Navy Yard more than 25 years ago—recommends that the key to the asbestos cases is their aggregation. With a single judge, a class-action suit will be brought to the courts, centralizing all plaintiffs and defendants. An equitable plan could then be established for payment based on the needs of the victims.

Perspective

Since the beginning of the Industrial Revolution the contamination of the environment has steadily increased. In spite of this, the human population has not only survived but has grown greatly in numbers, and life expectancy has increased. This raises a fundamental question: What is the purpose of managing toxic waste and hazardous materials in our environment? Is it to control the environment or is it to determine how humans use the environment? There are those who believe that our society is wasteful, morally divided, and sustained only by polluting, resource-depleting, high technology. Their conclusion is that, no matter what the benefits are, the long-term risks are too great and amount to a basic violation of human rights. In contrast, there are those who argue that the problems of environmental contamination can be resolved by modern scientific technology, and that the future is most promising.

The problem thus is not only technological but also ethical. It is undeniable that regulatory standards and institutionalized structures are necessary for the protection of some measure of human health. However, what constitutes a hazard depends upon values on a scale of real possibilities. In essence, hazards are a response to value standards. Further hazards are not absolute. Consequently, many have adopted the uncritical assumption that risk is a normative concept certifying that everything can be classified either good or bad. As a result, a false antithesis has been established between risks and benefits—as if there were a way to have one without the other. In reality, in the decision process what is actually "at risk" is the possibility that the intended benefits from risk taking may not occur. If harm occurs it is an unwanted and unintended side effect. Thus risks and benefits are inseparable. To illustrate, if we are to have nature produce a perfect apple, there must be a control of insects. In controlling insects in order to achieve the perfect fruit, there may be some risk. Thus each qualitative benefit has a corresponding level of risk.

It is usually not recognized that there are degrees of harm, extending from trivial to disastrous. If a harm is trivial and prevents a greater harm, there is ethical justification for the trivial harm. In contrast, if the harm is so severe that human disaster results, then there is no ethical justification for this act. One of the greatest problems in determining the magnitude of harm to a human being is the incremental risk of exposure to a toxic substance over time. A single exposure or several exposures may cause little or no harm, while continuous exposure may result in great risk to human health. Under this condition there must be a restructuring of the prior patterns of benefits and risks. Only systematic risk accounting can provide accurate data for this type of situation. To illustrate, a single contact with a toxic chemical may result in no health risk, but continual contact may result in death.

There is an additional problem in determining benefits and risks, in that there are many naturally occurring toxics. In the creation of any policy for the use of toxic materials, the policymakers must initially take into account the wide variation of personal exposure to naturally occurring background sources. Human tolerance for natural toxicants demonstrates that increments from human-made applications can be kept well within the range of variance without inflicting unjustifiable harm on the general population or depriving humankind of possible benefits.

There is an ethical need for the formulation of principles essential for providing a sound basis for the proper management of hazardous material and toxic waste, management that will protect both the environment and the health of individuals. As a fundamental premise it must be recognized that such a policy cannot be guided by the philosophy that no risk is valid. Any policy will entail risks of some harm to the environment and individuals at some future time. A policy for the protection of health must be based on a method of assessing systematic risks so as to prevent the environment and human beings from experiencing basic harm deemed unjustifiable.

This policy is thus not based on the absence of risk but rather on how much risk is acceptable. There is an urgent need to reach a consensus on this fundamental question. Without an answer to this problem the basic rights to health protection and a quality environment may be lost in the obsession with hypothetical health effects from only a few toxic substances. This distorted viewpoint averts public attention from a host of more pressing problems such as preventable causes of malnutrition and diseases. Thus, a policy of social justice requires management of the potential sources of basic harm, that is, management that is proportional to actual, identifiable basic harms that human effort and time can reduce.

To implement such a policy, we need a full evaluation of the technology available for the reduction of each potential toxic hazard. Only then can standards be established to provide the greatest benefits for the most people. It must be remembered that risk assessment is not a utilitarian tool for placing a monetary value on human welfare. In reality, risk assessment maximizes the value placed on human welfare so that a finite amount of money can be used to maximize the greatest benefits to human welfare and the quality of the environment.

2

Chronology

THE POLLUTION OF THE ENVIRONMENT by toxic waste and hazardous material has a long history. There are many types of hazardous waste that cause a deterioration of the environment and affect human health. The following listing of critical dates and their significance may help further our understanding of one of the major problems facing humankind today.

Roots of Environmentalism and Human Welfare

1840s   It is recognized that impure public water can spread typhoid through a community.

1849   Dr. John Snow in London recognizes that cholera outbreaks are concentrated in areas of sewage-contaminated waters. Dr. Snow is an early environmentalist, attempting to prevent an environmentally caused disease.

1874   G. P. March indicates that human activity will eventually affect the earth's environment.

1913   Herbert Quick writes in his book, *The Good Ship Earth,* that life on earth depends on a closed system of living things and physical objects.

1927   Edward A. Ross is one of the early writers on the effects of increasing population of the world on the earth's environment.

1950s   John Maddox, author of *The Doomsday Syndrome,* traces the modern environmental movement to widespread fear resulting from the radioactivity released by the atmospheric nuclear weapons tests of the 1950s. At this time concern begins to be expressed not only by scientists, but also by the general public, about the effects of radioactivity on the environment in general and on human health in particular.

Pesticides

1000 B.C.   Sulfur is known to control plant diseases. Homer mentions sulfur as a fumigant in his writings.

300 B.C.   Theophrastus describes many plant diseases known in modern times as rust, scab, rot, and scorch.

79 A.D.   Pliny advocates the use of arsenic as an insecticide.

1500s   The Chinese use arsenic compounds to control insects in agriculture.

1600s   The first natural insecticide, nicotine, extracted from tobacco, is used to control the lace bug and plum curculio.

1841   In England a solution compound of tobacco, sulfur, and unslaked lime is prepared to control insects and fungi.

1850s   Two important natural insecticides are introduced—rotenone, from the roots of the derris plant, and pyrethrum, from the flower heads of a species of chrysanthemum. These are still widely used insecticides.

              Soap is used to kill aphids and sulfur is used as a fungicide on peach trees. A mixture of sulfur with lime to soften it, later called lime sulfur, becomes an inexpensive insecticide.

1867   Impure copper arsenate, Paris green, is introduced to control the Colorado beetle in the state of Mississippi. By 1900 Paris green is used so extensively as an insecticide that it causes the introduction of the first state legislation controlling the use of insecticides in the United States.

1874   DDT, the chemical dichloro-diphenyl-trichloro-ethane, is first synthesized by a German chemist.

1886   Cyanide, generally as hydrogen cyanide gas, begins to be used against scale insects on citrus trees in California. Tents are placed over the trees and the hydrogen cyanide is generated inside. Initially this proves effective, but failures begin to occur as the scale insects develop resistant strains. This is the first reported example of resistant strains to an insecticide.

1896   A French farmer applying a Bordeaux mixture to vineyards notices that only the leaves of charlock die. This incident is the first one that records the concept of a selective herbicide.

1900   It is discovered about this time that when iron sulfate is sprayed on a field of cereals and weeds, only the weeds are killed. In the next decade many other inorganic compounds, such as copper sulfate, ammonium sulfate, and sulfuric acid, are found to exhibit selective herbicidal qualities at suitable concentrations.

1912   W. C. Piver develops calcium arsenate as a replacement for Paris green and lead arsenate. Calcium arsenate provides a major control of the cotton boll weevil.

1920s   The extensive application of arsenical insecticides causes widespread public concern because treated fruits and vegetables are sometimes contaminated with the poisonous residues.

             A search begins to replace inorganic chemicals with less dangerous organic pesticides such as tar, petroleum oils, and dinitro-o-cresol. The last of these is patented in 1933 as a selective herbicide (Sinox) to control weeds in grain fields.

1930s   This is the beginning of the modern era of synthetic organic pesticides. Important examples include the alkyl thiocyanate insecticides (1930); salicylanilide, or shirlan (1931); the first organic fungicide, dithiocarbamate fungicide (1934); and a spray to control pathogenic fungi, chorahil (1938). Also in this decade, the extreme toxicity of PCBs is recognized and concern begins as to whether it is a safe insecticide when sprayed on human food.

1939   Paul Müller discovers that DDT can control insects. After successful field tests in Switzerland against the Colorado potato beetle, commercial production begins in 1943. In 1949 Dr. Müller is awarded the Nobel Prize in medicine and physiology for this work.

1940–   DDT controls louse-borne typhus and malaria-carrying mosquitoes
1960s   on a worldwide scale. It is an effective insecticide in World War II because it permits military operations in the tropics, where the danger from epidemics is great.

1942   Benzene hexachloride, first prepared by the English chemist Michael Faraday in 1825, is recognized as an insecticide.

1948   Aldrin and dieldrin are introduced as pesticides.

1957   The U.S. Department of Agriculture begins spraying to eradicate the gypsy moth in the northeastern United States.

May 23. The Fire Ant Eradication Act is signed into law by President Eisenhower. Spraying to eradicate the fire ant in the southeastern United States begins immediately.

1960s   Pesticides are added to nuclear fallout in discussions concerning the contamination of the environment.

Agent Orange, containing dioxin, is used to control tropical vegetation in Vietnam.

1962   Rachel Carson's *Silent Spring* is published, bringing to the attention of the general public the reckless attempt to control our environment by the use of chemicals that poison not only the insects against which they are directed, but also birds, fish, the earth, and potentially humans.

1970   The U.S. Environmental Protection Agency is given responsibility for administration of federal pesticide laws.

1971–   In 1971 the Environmental Protection Agency begins its review of
1978   chlordane to determine if it is a carcinogen. In 1974 the review of heptachlor begins. In October 1974 the Environmental Defense Fund petitions the EPA to ban chlordane and heptachlor on the grounds that they "pose an imminent health hazard to man." In November 1974 EPA Administrator Russell E. Train issues a "Notice of Intent" to cancel most uses of heptachlor and chlordane. In December 1975 the EPA announces an immediate temporary ban on most uses of these chemicals. In March 1978 the EPA places a permanent ban on most uses of chlordane by 1980, and heptachlor by July 1983.

1972   June 14. William Ruckelshaus, administrator of the Environmental Protection Agency, bans the use of DDT with the exception of uses related to essential public health purposes. His decision is based on

1972
cont. the belief that the chemical poses an unacceptable risk, not only to wildlife and the environment, but to humans as a carcinogen.

1974 August 2. The Environmental Protection Agency suspends the manufacture of insecticides aldrin and dieldrin on the grounds that they pose "an unacceptably high cancer risk."

1978 A class-action suit is filed by 40,000 veterans and their families who may have been contaminated by Agent Orange spraying in Vietnam.

1980 July. The Environmental Protection Agency places a ban on most uses of lindane as an insecticide.

1982 The Environmental Protection Agency restricts most uses of the insecticide toxaphene, while permitting the livestock dipping program to continue.

1984 May 6. A court trial against chemical companies producing Agent Orange is set for May 7, 1984, in Brooklyn. Before the trial begins, an out-of-court settlement is reached totaling $180 million for Vietnam veterans.

1990 Although the use of a number of insecticides has been restricted, there are hundreds still on the commercial market to be tested by the Environmental Protection Agency as to their environmental and human impact.

Major Toxic Waste Sites

1950s– Niagara Falls, New York. Hooker Chemical Company dumps mil-
1970s lions of pounds of hazardous wastes in the local municipal dumps.

1960s– Lathrop, California. The Occidental Chemical Company discharges
1970s thousands of gallons of pesticide wastes into the ground on the company's site. The waste products placed in lagoons are allowed to percolate into the extremely permeable soil, threatening the area's drinking and irrigation water.

1970s Valley of the Drums, Shepherdsville, Kentucky, contains hundreds of hazardous wastes. Thousands of barrels are stacked illegally in a hauler's backyard. By the late 1970s these barrels are in a seriously deteriorating state; some have burst and spilled their contents on the ground.

1970s
cont.
East Texas. Waste oil contaminated with toxic chemicals is spread on nine roads because of the negligence of the waste disposal company.

Hardeman County, Tennessee. Liquid hazardous waste barrels are rolled into trenches and tank wagons are discharged directly into pits on reclaimed land. There is evidence chemicals entered into the water supply.

Bayou Sorrel, Louisiana. Millions of gallons of toxic wastes are dumped into open lagoons, filling the air with asphyxiating toxicant. One truck driver employed by a disposal company dies of hydrogen sul- fate asphyxiation as he empties chemicals into a waste pit.

Saltville, Virginia. An alkali processing plant operating between 1895 and 1972 leaves large deposits of mercury-contaminated rocks that pollute the North Fork of the Holston River.

Pickens, South Carolina. An electric company dumps dozens of capacitors and transformers leaking high levels of PCBs into a watershed, contaminating local drinking water.

Elizabeth, New Jersey. A chemical disposal site contains over 40,000 barrels of hazardous wastes. At least 100 pounds of picric acid, a powerful explosive, are also found stored on the site. Thousands of barrels of highly toxic, explosive, and flammable materials are unsafely stored within a few feet of a waste incinerator and a few feet from a public road.

1976–1983
Love Canal, Niagara Falls, New York. Hooker Chemical Company deposits in three sites an estimated 353 million pounds of industrial chemical waste, including TCP (which is often contaminated with dioxin) and lindane, a highly toxic pesticide product.

1982–1983
Times Beach, Missouri. This area is contaminated by dioxin. In December 1982, after the area is flooded, it is feared that dioxin has spread throughout the town. The EPA spends $33 million to remove the residents from Times Beach.

1990
July. The Environmental Protection Agency declares Love Canal, Niagara Falls, New York, free of contaminants, and begins to sell abandoned houses in the process of resettlement.

August. The Environmental Protection Agency announces plans to clean up Times Beach, Missouri. An incinerator will be built to burn toxic wastes. Resettlement of the area is planned for the future.

Chemical Accidents and Explosions of Hazardous Materials

1913 March 7. An explosion of dynamite in Baltimore, Maryland, harbor kills 55.

1917 April 10. An explosion in a munitions plant in Eddystone, Pennsylvania, kills 133.

December 6. Explosions of ammunition in ships in Halifax, Nova Scotia, harbor kill 1,654.

1918 May 18. An explosion in a chemical plant in Oakdale, Pennsylvania, kills 193 workers.

October 4. An explosion in a shell plant in Morgan Station, New Jersey, kills 64.

1940 September 12. An explosion in the Hercules Powder plant in Kenvil, New Jersey, kills 55.

1948 June 20. An explosion of chemicals in the Farben works in Ludwigshafen, West Germany, kills 1,984.

1956 August 7. Dynamite trucks explode in Cali, Colombia, killing 1,100.

1960 March 4. A Belgian munitions ship explodes in Havana, Cuba, harbor, killing 100.

1963 March 9. An explosion in a dynamite plant in South Africa kills 45.

August 13. An explosion in an explosives dump in Gauhiti, India, kills 32.

1966 October 13. An explosion in a La Salle, Quebec, chemical plant kills 11.

1967 February 17. An explosion in a chemical plant in Hawthorne, New Jersey, kills 11.

1973 February 10. A liquefied gas tank explodes on Staten Island, New York, killing 40.

1976 April 13. A munitions works in Lapua, Finland, experiences an accident, killing 46.

1976
cont.
July. An accident in a chemical plant at Sweso, Italy, releases a cloud of highly toxic dioxin (tetrachlorodibenzo-p-dioxin) into the atmosphere. Because of a stable air mass the pollutant remains in the area about three weeks, forcing the evacuation of 700 people, at least 500 of whom exhibit symptoms of poisoning. About 600 animals are poisoned and have to be destroyed, and contaminated crops are burned. Medical experts recommend that all residents of the area receive periodic medical examinations for the rest of their lives.

1978
July 11. A propylene tank truck explodes at a Spanish coastal campsite, killing 150.

1984
December 20. In Bhopal, India, water leaks into a tank containing the chemical compound isocyanate methyl at the Union Carbide pesticide plant. The ensuing intense reaction creates a lethal gas that erupts, flowing over the town and causing the greatest industrial disaster known in history. The number of deaths totals at least 2,850 and those seriously afflicted number possibly 40,000.

1985
June 25. A fireworks factory at Hallett, Oklahoma, explodes, killing 21.

Major Oil Spills

Besides the major oil spills, there have been hundreds of small accidental spills on land and in the oceans. In addition, there has been deliberate dumping of oil in the oceans and on the land as a waste product.

1967
March 18. The tanker *Torrey Canyon* grounds on the Seven Stones Shoal off the coast of Cornwall, England, spilling 830,000 barrels of Kuwaiti oil into the sea. This is the first major tanker accident.

October 15. A dragging ship anchor punctures a pipeline in the Mississippi at West Delta, Louisiana, spilling 160,000 barrels of oil.

1968
March 3. The tanker *Ocean Eagle* grounds at the entrance to the San Juan, Puerto Rico, harbor, spilling its cargo into the harbor.

June 13. The tanker *World Glory* experiences hull failure off South Africa and spills nearly 320,000 barrels of oil in the south Atlantic Ocean.

1969   January 28. An oil spill occurs in the coast off of Santa Barbara, California, because of lack of control of underground oil pressure during drilling operations.

November 4. A storage tank ruptures in Sewaren, New Jersey, spilling 200,000 barrels of oil on land surfaces.

November 5. The tanker *Keo* experiences hull failure off the Massachusetts coast and spills 210,000 barrels of oil into the ocean.

1970   March 20. The tanker *Othello* experiences a collision in Tralhavet Bay, Sweden, spilling more than 420,000 barrels of oil.

1971   November 30. An oil tanker off the coast of Japan breaks in half, dumping 150,000 barrels of oil into the Pacific Ocean.

1972   December 19. The tanker *Sea Star* experiences a collision in the Gulf of Oman and spills 800,000 barrels of oil.

1976   May 12. The tanker *Urqusola* strikes an underwater obstruction entering the harbor of La Coruna, Spain, and spills 700,000 barrels of oil.

December 15. The tanker *Argu Merchant* grounds off Nantucket, Massachusetts, spilling over 180,000 barrels of oil.

1977   February 28. The tanker *Hawaiian Patriot* catches fire and has to dump 690,000 barrels of oil in the northern Pacific Ocean.

April 22. The Ekofisk oil field in the North Sea experiences a well blowout; 195,000 barrels of oil flow into the sea.

December 16. The oil tankers *Ven Oil* and *Ven Pet* collide off Port Elizabeth, South Africa, spilling some 210,000 barrels of oil.

1978   March 16. *Amoco Cadiz,* carrying a cargo of 1,540,000 barrels of light Arabian crude oil, experiences rudder failure near the English Channel and grounds on the Brittany coast, spilling its entire cargo.

1979   June 3. Idoc's number one well experiences a blowout in the southern Gulf of Mexico, spilling 86,000 barrels of oil.

July 19. The tankers *Atlantic Empress* and *Aegean Captain* collide off Trinidad and Tobago, depositing 2.1 million barrels of oil in the sea.

November 1. The tanker *Burmah Agate* collides with another vessel in Galveston Bay, Texas, spilling about 280,000 barrels of oil.

1983   February. A blowout of a well in the Newraz oil field in the North Sea deposits 4.2 million barrels of oil in its vicinity.

August 6. The Spanish supertanker *Castillo de Bellver,* laden with 1.75 million barrels of Persian Gulf oil, bursts into flame about 80 miles northeast of Capetown, South Africa. The tanker breaks apart, spilling its oil into the ocean.

1988   January 2. At Ashland, Pennsylvania, on the Monongahela River, an oil bulk storage facility fails, spilling about 90,000 barrels of no. 2 fuel oil onto the river's edge. The fuel oil contaminates the Monongahela and Ohio rivers to the Ohio-Pennsylvania border.

1989   March 24. The *Exxon Valdez* tanker grounds on Bligh Reef in Prince William Sound near Valdez, Alaska, spilling nearly 250,000 barrels of crude oil and creating the worst oil spill in the history of the United States.

Laws and Regulations

1947   Federal Insecticide, Fungicide and Rodenticide Act of 1947, Public Law 1033.

1965   Solid Waste Disposal Act of 1965, Public Law 89-272.

1970   Hazardous Materials Transportation Control Act of 1970, Public Law 91-458, Title III.

1972   Federal Environmental Pesticide Control Act of 1972, Public Law 92-516; amends the Federal Insecticide, Fungicide and Rodenticide Act of 1947.

1976   Resource Conservation and Recovery Act of 1976, Public Law 94-580; amendment of the Solid Waste Disposal Act of 1965.

Toxic Substances Control Act of 1976, Public Law 94-469, and as amended 1983, Public Law 98-80.

1980   Solid Waste Disposal Act Amendments of 1980, Public Law 96-482.

Hazardous Substance Response Revenue Act of 1980, Public Law 96-510, Title II.

Asbestos School Hazard Detection and Control Act of 1980, Public Law 96-270.

1980   Comprehensive Environmental Response, Compensation and Lia-
cont.   bility Act of 1980, Public Law 96-510.

1981   Consumer Product Safety Amendments of 1981, Public Law 97-35,
        Title XII Omnibus Budget Reconciliation Act of 1981.

1984   Asbestos School Hazard Abatement Act of 1984, Public Law 98-377,
        Education for Economic Security Act of 1984, Title IV.

1986   Emergency Planning and Community Right-To-Know Act of 1986,
        Public Law 94-499, Superfund Amendments and Reauthorization
        Act of 1986, Title III.

        Superfund Amendments and Reauthorization Act of 1986, Public
        Law 99-499.

1988   School Asbestos Management Plans Act of 1988, Public Law
        100-368.

        Lead Contamination Control Act of 1988, Public Law 100-572.

        Asbestos Information Act of 1988, Public Law 100-577.

        Pesticide Monitoring Improvements Act of 1988, Public Law 100-
        418, Omnibus Trade and Competitiveness Act of 1988, Title II,
        Agricultural Trade, Subtitle G Pesticide Monitoring Improvements.

3

Laws and Regulations

IN THE PAST HALF CENTURY it has become evident that humans are polluting the environment. This recognition has led to a series of laws at both federal and state levels. Many of the initial laws concentrated on the hazards of radioactive waste and the pollution of air and water. Of the hazardous and toxic pollutants of the earth, pesticides and herbicides received initial attention. In the 1960s such laws as the Solid Waste Disposal Act began to be directed toward the control of hazardous and toxic wastes in the environment.

In the 1970s the Resource Conservation and Recovery Act of 1976 and the Comprehensive Environmental Response, Compensation and Liability Act of 1980 (Superfund) established the nation's basic hazardous waste management system. At the same time, they provided the complementary authority to encourage the conservation and recovery of valuable materials and energy. A number of additional laws have been enacted to develop specific aspects of the comprehensive initiative.

Basic Hazardous and Toxic Waste Legislation

Solid Waste Disposal Act of 1965, Public Law 89-272 Hazardous and Solid Waste Amendment of 1984, Public Law 98-616

The purpose of this act was to initiate and accelerate a national research and development program for new and improved methods for economic solid waste disposal. These studies were to include conservation of natural resources by reducing the amount of waste and unsalvageable materials and by recovery and utilization of potential resources in solid waste. The term *solid waste* was defined as garbage, refuse, and other discarded solid materials, including waste from industrial, commercial, agricultural, and community activities.

Congress based this act on the following findings:

1. That the continuing technological progress and improvement in methods of manufacture, packaging, and marketing of consumer products has resulted in an ever-mounting increase, and in a change in the characteristics, of the mass of material discarded by the purchase of such products.
2. That the economic and population growth of our nation, and the improvements in the standard of living enjoyed by our population, have required increased industrial production to meet our needs.
3. That the continuing concentration of our population in expanding metropolitan and other urban areas has presented these communities with serious financial, management, intergovernmental, and technical problems in the disposal of solid wastes resulting from the industrial, commercial, domestic, and other activities carried on in such areas.
4. That inefficient and improper methods of disposal of solid wastes result in scenic blights, create serious hazards to the public health, including pollution of air and water resources, accident hazards, and increase in rodent and insect vectors of disease, have an adverse effect on land values, create public nuisances, and otherwise interfere with community life and development.
5. That the failure or inability to salvage and reuse such materials economically results in the unnecessary waste and depletion of our natural resources.
6. That while the disposal of solid waste should remain primarily a function of state, regional and local agencies, the problem has become national in scope and therefore financial and technical assistance must begin at the national level.

The provisions of the act were to be carried out by dissemination of information acquired through research studies. Grants were to be given to public or private agencies for research, training projects, screenings, and demonstrations.

The 1984 amendment elaborated such aspects of the act as land disposal of hazardous waste, minimum technological requirements, burning and blending of hazardous waste, groundwater monitoring, financial responsibility for corrective action, waste minimization facilities, mandatory inspections and federal enforcement. A National Groundwater Commission was also established.

National Environmental Policy Act of 1969, Public Law 91-190

The purpose of this act was to establish a national policy for the environment. As stated it was:

1. To develop a national policy which will encourage productive and enjoyable harmony between man and his environment.
2. To promote efforts which will provide or eliminate damage to the environment and biosphere and stimulate the health and welfare of man.
3. To enrich the understanding of the ecological system and natural resources important to the nation.
4. To establish a Council on Environmental Quality.

This act also established the basic policy for the environmental acts enacted in the 1970s and 1980s. In the establishment of a natural environmental policy, Congress,

> recognizing the profound impact of man's activity on the interrelations of all components of the natural environment, particularly the profound influence of population growth, high-density urbanization, industrial expansion, resource exploration, and new and expanding technological advances and further recognizing the initial importance of restoring and maintaining environmental quality to the overall welfare and development of man, declares that it is the continuing policy of the Federal Government, in cooperation with the state and local governments, and other concerned public and private organizations, to use all practicable means and measures, including financial and technical assistance, in a manner calculated to foster and promote the general welfare, to create and maintain conditions under which man and nature can exist in productive harmony, and fulfill the social, economic, and other requirements of present and future generations of Americans.

In order to carry out this policy, it is the responsibility of the federal government to implement and coordinate federal plans, functions, programs, and resources so that the nation may do the following:

1. Fulfill the responsibilities of each generation as trustee of the environment for succeeding generations.
2. Assume for all Americans safe, healthful, productive, and aesthetically and culturally pleasing surroundings.
3. Attain the widest range of beneficial uses of the environment without degradation, risk of health or safety, or other undesirable and unintended consequences.

4. Preserve important historic, cultural, and natural aspects of our national heritage, and maintain, whenever possible, an environment which supports diversity and variety of individual choice.
5. Achieve a balance between population and resource use which will permit high standards of living and a wide sharing of life's amenities.
6. Enhance the quality of renewable resources and approach the maximum attainable recycling of depletable resources.

The Congress authorized that all agencies of the government take the following steps:

1. Utilize a systematic, interdisciplinary approach which insures the integrated use of the natural and social sciences and the environmental design acts.
2. Identify and develop methods and procedures in order to insure that presently unquantified environmental amenities and values may be given appropriate consideration in decisionmaking.
3. Include in every recommendation, (1) the environmental impact of the proposed action, (2) any adverse environmental effects, (3) alternatives to the proposed action, (4) relationships between local short-term use of man's environment and the maintenance and enhancement of long-term productivity, and (5) any irreversible and irretrievable commitments of resources.
4. Study, develop, and describe appropriate alternatives to recommended courses of action.
5. Recognize the worldwide and long-range character of environmental problems.
6. Make available to states, counties, municipalities, institutions, and individuals advice and information useful in restoring, maintaining, and enhancing the quality of the environment.
7. Initiate and utilize ecological information in the planning and development of resource-oriented projects.

The provisions of this act were incorporated in the Reorganization Plans Nos. 3 and 4 submitted to Congress on July 9, 1970, by President Richard Nixon in the establishment of the Environmental Protection Agency and the National Oceanic and Atmospheric Administration.

Resource Conservation and Recovery Act of 1976, Public Law 94-580
Amendment of the Solid Waste Disposal Act of 1965, Public Law 89-272

The Solid Waste Disposal Act of 1965 emphasized the recovery of waste materials. The Resource Conservation and Recovery Act of 1976 continued these efforts but recognized that wastes were frequently hazardous to the environment and health. Congress found the following in respect to the environment and health:

1. Although land is too valuable a national resource to be needlessly polluted by discarded materials, most solid waste is disposed of on land in open dumps and sanitary landfills.
2. Disposal of solid waste and hazardous waste in or on the land without careful planning and management can present a danger to human health and the environment.
3. As a result of the Clean Air Act, the Water Pollution Control Act, and other Federal and State laws respecting public health and the environment, greater amounts of solid waste (in the form of sludge and other pollution treatment residues) have been created. Similarly, inadequate and environmentally unsound practices for the disposal or use of solid wastes have created greater amounts of air and water pollution and other problems for the environment and for health.
4. Open dumping is particularly harmful to health, contaminates drinking water from underground and surface supplies, and pollutes the air and land.
5. Hazardous waste presents, in addition to the problems associated with non-hazardous solid waste, special dangers to health and requires a greater degree of regulation than does non-hazardous solid waste.
6. Alternatives to existing methods of waste disposal must be developed since many of the cities in the United States will be running out of suitable solid waste disposal sites within five years unless immediate action is taken.

In the act, Congress also recognized the importance of the recovery and conservation of materials from solid waste and the use of solid waste to produce energy.

The objectives of the act were to promote the protection of health and the environment and to ensure valuable materials and energy resources by doing the following:

1. Providing technical and financial assistance to state, local governments and interstate agencies for the development of solid waste management plans.
2. Providing training grants in occupations involving the design, operation, and maintenance of solid waste disposal systems.
3. Prohibiting future open dumping on the land and requiring the conversion of existing dumps to facilities which do not pose a danger to the environment or to health.
4. Regulating the treatment, storage, transportation, and disposal of hazardous wastes which have adverse effects on health and the environment.
5. Providing for the promulgation of guidelines for solid waste collection, transport, separation, recovery, and disposal practices and systems.

6. Promoting a national research and development program for improved solid waste management and resource conservation techniques, more effective organizational arrangements, and new and improved methods for collection, separation, and recovery, and recycling of solid wastes and environmentally safe disposal of nonrecoverable residues.

7. Promoting the demonstration, construction, and application of solid waste management, resource recovery, and resource conservation systems which preserve and enhance the quality of air, water and land resources.

8. Establishing a cooperative effort among the Federal, state and local governments and private enterprises in order to recover valuable materials and energy from solid waste.

This act defined a number of terms:

1. The term "hazardous waste" means a solid waste, combination of solid wastes, which because of its quantity, concentration, or physical, clinical or infectious characteristics may: (A) cause, or significantly contribute to an increase in mortality, or an increase in serious irreversible, or incapacitating reversible illness, (B) or pose a substantial present or potential hazard to human health or the environment when improperly treated, stored, transported, or disposal of, or otherwise managed.

2. "Hazardous waste generation" means the act or process of producing hazardous waste.

3. "Hazardous waste management" means the systematic control of the collection, source separation, storage, transportation, processing, treatment, recovery, and disposal of hazardous wastes.

4. "Open dump" means a site for the disposal of solid waste which is not a sanitary landfill.

5. A "sanitary landfill" is a site where there is no reasonable probability of adverse effects on health or the environment from the disposal of solid waste.

This act requires that an Office of Solid Waste be established within the Environmental Protection Agency. This office has the following responsibilities:

1. Prescribe in consultation with federal, state, and regional authorities necessary regulation to carry out the functions of the Act.

2. Consult with other agencies for the purpose of the exchange of information.

3. Provide technical and financial assistance to states and regional agencies in development of solid waste plans and management strategies.

4. Consult with representatives from science, industry, agriculture, labor, environmental protection and consumer organizations.

5. Utilize information from other agencies such as the National Bureau of Standards and the National Bureau of the Census.

A major feature of the act is the management of hazardous waste (Subtitle C). This requires that criteria be established for identifying the characteristics of hazardous wastes. Factors to be considered are toxicity, persistence, degradability in nature, potential for accumulation in tissue, and such related factors as flammability, corrosiveness, and other hazardous characteristics.

The act requires that standards be established. Such standards must include the following:

1. Recordkeeping practices that accurately identify the quantities of such hazardous waste generated, the constituents which are significant in quantity that they provide a potential harm to human health or the environment including the disposition of such waste.
2. Proper labeling for any containers used for the storage, transport, or disposal of such hazardous waste.
3. Use of appropriate containers for such hazardous waste.
4. Furnishing information on the general chemical composition of hazardous waste to persons transporting, treating, storing or disposing of wastes.
5. Use of a manifest system to assure that all hazardous waste generated is designated for treatment, storage, or disposal in treatment, storage, or disposal facilities for which a permit has been issued.
6. Submission of reports to the EPA of the quantities of hazardous waste identified during a particular time period and the means of its disposal.

The act also required that the following standards be established for the transportation of hazardous materials:

1. Recordkeeping concerning such hazardous materials transported, and their source and delivery points.
2. Transportation of such wastes only if properly labeled.
3. Compliance with the manifest system listed above.
4. Transportation of all hazardous waste to the hazardous waste treatment, storage, or disposal facilities which the shipper designates on the manifest form to be a facility holding a permit.

In order to enforce the management of the act, the EPA has the responsibility of inspection of any site that generates, stores, treats, transports, disposes of, or otherwise handles hazardous wastes. Such an officer has the following authority:

1. To enter at reasonable times any establishment or other place maintained by any person where hazardous wastes are generated, stored, treated or disposed of.
2. To inspect and obtain samples from any person where hazardous wastes are generated, stored, treated or disposed of.

If the provisions of this act are not met, federal enforcement stipulates a time period, usually 30 days, within which violators must comply with the regulations. If the violation is not corrected, a penalty of not more than $25,000 for each day of noncompliance may be imposed, and the EPA may suspend or evoke the permit.

This act, in Subtitle D, provides for state and regional solid waste plans. The objectives of this subtitle are to assist in developing and encouraging methods of solid waste that are environmentally sound and that maximize the utilization of valuable resources and to encourage resource conservation.

The federal guidelines for solid waste plans include encouraging and facilitating the development of regional planning for solid waste management. Such guidelines consider the following:

1. The size and location of areas which should be included.
2. The volume of solid waste which should be included.
3. The available means of coordinating regional planning with other related regional planning and for coordination of such regional planning into the state plan.

The state plan guidelines consider these characteristics:

1. The varying regional, geologic, hydrologic, climatic, and other circumstances under which different solid waste practices are required in order to insure the reasonable protection of the quality of the ground and surface waters from leaked contamination, the quality of the surface water from surface runoff contamination, and the reasonable protection of ambient air quality.
2. Characteristics and conditions of collection, storage, processing, and disposal operating methods, techniques and practices, and the location of facilities where such operating methods, techniques and practices are conducted.
3. Methods for closing or upgrading open dumps for purposes of eliminating potential health hazards.
4. Population, density, distribution and projected growth.
5. Geographic, geologic, climatic, and hydrologic characteristics.
6. The type and location of transportation.
7. The profile of industry.
8. The constituents and generation rates of waste.
9. The political, economic, organizational, financial, and management problems affecting comprehensive solid waste management.
10. Types of resource recovery facilities and resource conservation systems which are appropriate.
11. Available new and additional sources for recovered materials.

The act recognizes that open dumps present a health and environmental hazard. As a response, any solid waste management practice or disposal of solid waste as hazardous waste that consists of open dumping is prohibited. Open dumps must be upgraded to comply with the law.

In order to implement the act, procedures for development and implementation of state plans were stated, including types of federal assistance. Special attention was given to small rural communities. Assistance is available:

1. To any municipality or county which could not feasibly be included in a solid waste management system or facility serving an urbanized, multijurisdictional area because of its distance from such systems.
2. Where existing or planned solid waste management services or facilities are unavailable or insufficient to comply with this Act.
3. For systems which are certified by the State to be consistent with any plans or programs established under any state or area wide planning process.

The act is also concerned with the recovery of resources. In Subtitle E the duties and responsibilities of the secretary of commerce are outlined as to resource recovery. These include the following:

1. Accurate specifications for recovered materials.
2. Stimulation of development of markets for recovered materials.
3. Promotion of proven technology.
4. A forum for the exchange of technical and economic data relating to resource recovery facilities.

The final subtitle of the act, Subtitle H, provides guidance for research, development, demonstration, and information. These include studies relating to the following:

1. Any adverse health and welfare effects of the release into the environment of material present in solid waste and methods to eliminate such effects.
2. The planning, implementation, and operation of resource recovery and resource conservation systems and hazardous waste management systems, including the marketing of recovered resources.
3. The operation and financing of solid waste disposal programs.
4. The reduction of the amount of such waste and unsalvageable waste materials.
5. The development and applications of new and improved methods of collecting and disposing of solid waste and processing and recovering materials and energy from solid wastes.
6. The identification of solid waste components and potential materials and energy recoverable from such waste components.
7. Improvement in land disposal practices for solid waste which may reduce the adverse environmental effects.

8. Methods for the sound disposal of, or recovery of resources.

9. Methods of hazardous waste management, including methods of rendering such waste environmentally safe.

10. Any adverse effects on air quality which result from solid waste which is bound for purpose of disposal or energy recovery.

Comprehensive Environmental Response, Compensation and Liability Act of 1980, Public Law 96-510 (Superfund)

This act provides liability, compensation, cleanup, and emergency response for hazardous substances released into the environment and the cleanup of inactive hazardous waste disposal sites.

This law gives authority to the president to act when any hazardous substance, pollutant, or contaminant is released into the environment. A plan must be devised for remedial action.

Hazardous Substance Releases, Liability, Compensation Act of 1980, Title I

Title I establishes a National Contingency Plan that develops procedures and standards for responding to release of hazardous substances, pollutants, or contaminants, including the following:

1. Methods for discovering and investigating facilities at which hazardous substances have been disposed of, or otherwise come to be located.

2. Methods for evaluating, including analyses of relative cost, and remedying any releases or threats of releases from facilities which pose substantial danger to the public health or the environment.

3. Methods and criteria for determining the appropriate extent of removal, remedy, and other measures authorized by this Act.

4. Appropriate roles and responsibilities for the Federal, State and local governments and for interstate and nongovernmental entities in implementing the plan.

5. Provisions for identification, procurement, maintenance, and storage of response equipment and supplies.

6. A method for and assignment of responsibility for reporting the existence of such facilities which may be located on federally owned or controlled properties and any releases of hazardous substances from such facilities.

7. Means of assuring that remedial action measures are cost effective over the period of potential exposure to the hazardous substances or contaminated materials.

8. Criteria for determining priorities among releases or threatened releases throughout the United States for the purpose of taking remedial action, and to the extent practicable taking into account the potential urgency of such action for the purpose of taking remedial action. Criteria and priorities shall be based upon the relative risk or danger to public health or welfare or the environment.

9. The President shall list as part of the plan national priorities among the known releases or threatened releases throughout the United States and shall revise the list no less often than annually.

In order to protect the health of individuals, the Agency for Toxic Substances and Disease Registry was established within the Public Health Service. This agency cooperates with other health organizations of the government, and has the following responsibilities as well:

1. With the cooperation with the States establishes and maintains a national registry of serious diseases and illnesses and a national registry of persons exposed to toxic substances.

2. Establishes and maintains an inventory of literature, research, and studies of the health effects of toxic substances.

3. With cooperation with the States, and other agencies of the Federal government, establishes and maintains a complete listing of areas closed to the public or otherwise restricted in use because of toxic substance contamination.

4. In case of public health emergencies caused or believed to be caused by exposure to toxic substances, provides medical care and testing to exposed individuals, including tissue sampling, chromosomal testing, epidemiological studies and other assistance.

5. Conducts periodic survey and screening programs to determine relationships between exposure to toxic substances and illness.

A major feature of this law is the stipulation of financial responsibility. All manufacturers of hazardous substances must maintain evidence of financial responsibility consistent with the degree and duration of risk associated with its production, transportation, treatment, storage, or disposal.

In order to implement the law, a fund (which has come to be known as the Superfund) was established to pay for the following:

1. The costs of assessing both short-term and long-term injury to, destruction of, or loss of any natural resources resulting from a release of a hazardous substance.

2. The cost of Federal or State efforts in the restoration, rehabilitation, or replacement or acquiring the equivalent of any natural resources injured, destroyed, or lost as a result of a release of a hazardous substance.

3. Subject to available funds, the costs of a program to identify, investigate, and take enforcement and abatement action against releases of hazardous substances.

4. The costs of epidemiologic studies, development and maintenance of a registry of persons exposed to hazardous substances to allow long-term effect studies.

5. Subject to available funds, the cost of providing equipment and similar overhead.

6. Subject to available funds, the costs of a program to protect the health and safety of employees involved in response to hazardous substance releases.

Hazardous Substance Response Revenue Act of 1980, Public Law 96-510, Title II

Title II of the act provides the means to impose taxes, including a tax on petroleum and certain chemicals. A trust fund, known as the Hazardous Substance Trust Fund, was established. The fund is used for costs incurred in implementing the Comprehensive Environmental Response, Compensation and Liability Act of 1980 (CERCLA).

A special feature of the act is the imposition of a tax on hazardous waste. The 1980 tax was $2.13 per dry-weight ton of hazardous waste. Monies generated are to be used to clean up hazardous waste problems.

Superfund Amendments and Reauthorization Act of 1986, Public Law 99-499

The Superfund Amendments strengthened the 1980 Comprehensive Environmental Response, Compensation and Liability Act. A major addition defined the phrase "pollutants and contaminants" to:

> Include, but is not limited to, any element, substance, compound, or mixture including disease-causing agents, which after release into the environment and upon exposure, ingestion, inhalation or assimilation into any organism, either directly from the environment or indirectly by ingestion through food chains, will or may reasonably be anticipated to cause death, disease, behavioral abnormalities, cancer, genetic mutations, physiological malfunctions, or physical deformations, in such organisms as their offspring.

Title I of the Superfund Amendments provides an elaboration of the provision relating primarily to response and liability. These include response authorities, national contingency plan, reimbursement, liability, penalties, use of funds, health-related authorities, claim procedure, public participation, cleanup standards, regulations, and relativity to other laws. Title II considers such aspects as hazardous materials transportation, citizens suits, Indian tribes, research, development and demonstration, pollution reliability insurance, and the Department of Defense environmental restoration program. Title III consists of the Emergency Planning and Community Right-To-Know Act of 1986, and Title IV is the Radon Gas and Indoor Air Quality Research Act of 1986. Title V is cited as the Superfund Revenue Act of 1986.

Budget Reconciliation Act of 1986, Public Law 99-509, Part IV, Tax on Petroleum and Oil Spill Liability Trust Fund

This legislation provides for taxation to defray the costs of the Superfund and oil spills. The funds are secured through an environmental tax on petroleum. The Hazardous Substance Superfund financing rate is 11.7 cents a barrel; and for the Oil Spill Liability Trust Fund, 1.3 cents per barrel.

States' Statutory Authority

Since the passage of the Comprehensive Environmental Response, Compensation and Liability Act, many states have enacted laws and developed programs with authorities and capabilities similar to the federal Superfund program. Of the states' statutory authorities and provisions, 37 have full fund and enforcement capabilities in hazardous waste cleanup statutes, 7 have limited fund capabilities (e.g., limited to emergency response and CERCLA match), 9 contain enforcement authorities in separate statutes, and 1 has no enforcement capabilities.

There are 20 states that have statutory provisions for a National Priorities List (NPL), 15 states that report citizen suit provisions, and 11 states that provide some type of victim compensation (Table 3.1).

Supplementary Hazardous and Toxic Legislation

Toxic Substances Control Act of 1976, Public Law 94-469, and as Amended, 1983, Public Law 98-80

In order to justify the passage of the Toxic Substances Control Act, Congress found the following:

TABLE 3.1
Statutory Authorities and Provisions

STATE	STATUTE	ENFORCEMENT AUTHORITIES	CLEANUP FUND	PRIORITY LIST	VICTIM COMPENSATION	CITIZEN SUIT PROVISION
Region I						
Connecticut	Public Law 87-561	X				
	Emergency Spill Response Fund	X	X		WS	
	Uncontrolled Hazardous Substance Sites Act	X	X			
Maine	Oil and Hazardous Release Prevention and Response Act	X	X	X		
Massachusetts	Hazardous Waste Laws	X	X			X
New Hampshire	Hazardous Waste Management Act	X	X			
Rhode Island	Solid Waste Management Law Contingency Fund, Water Pollution Control Act	X	X		X(1)	
Vermont	State Superfund Act		X(3)		X(2)	X
Region II						
New Jersey	Spill Compensation and Control Act	X	X		X	
New York	Abandoned Sites Act of 1979	X		X		
Region III						
Delaware	Hazardous Waste Management Act	O	ER, CS	X		
Maryland	Code of Maryland Health-Environmental Article	X	X	X	WS	
Pennsylvania	Hazardous Sites Cleanup Act	X	X		WS	X
Virginia	Waste Management Act	X	X			
West Virginia	Hazardous Waste Energy Response Fund Act	O	X			
Region IV						
Alabama	Hazardous Substance Cleanup Fund	X	X			
Florida	Pollutant Discharge Prevention and Removal Act	X	X(5)		WS	
	Resource Recovery and Management Act	X(4)	X			

State	Legislation					
Georgia	Hazardous Waste Management Act	O	X			
Kentucky	Revise statutes annually	X	X			
Mississippi	Solid Waste Disposal Act of 1974	O	X			
North Carolina	Comprehensive Environmental Response Act	X	X			
South Carolina	Hazardous Waste Management Act	X	X			
Tennessee	Hazardous Waste Management Act of 1983	X	X			
Region V						
Illinois	Environmental Protection Act	X	X			
Indiana	Hazardous Waste Act	X	X			
Michigan	Environmental Response Act	O	X			
Minnesota	Environmental Response and Liability Act	X	X			
Ohio	Solid and Hazardous Waste Disposal Law	X	X			X
Wisconsin	Environmental Response Statute	X	X(6)			
Region VI						
Arkansas	Remedial Action Fund	X	X			
	Emergency Response Fund Act	X	X			
Louisiana	Environmental Quality Law	X	X			
New Mexico	Hazardous Waste Act	O	ER, CS			
Oklahoma	Controlled Industrial Waste Disposal Act	O	ER, CS			
Texas	Solid Waste Disposal Act	X	X(7)	X		
Region VII						
Iowa	Environmental Quality Act	X	X	X	WS	
Kansas	Environmental Response Act	X	X	X		X
Missouri	Hazardous Waste Management Act	X	X	X		
Nebraska	Environmental Protection Act	O(8)	X			
Region VIII						
Colorado	Hazardous Substance Response Fund	O	CS (9)			
Montana	Comprehensive Environmental Cleanup and Response Act	X	X	X		
North Dakota	Hazardous Waste Management Act	O	X(10)			
South Dakota	Regulated Substance Discharge Law	X	X			
Utah	Hazardous Substance Mitigation Act	X	X	X		
Wyoming	Environmental Quality Act	O	ER			

continued

ER = Emergency response and removal; CS = CERCLA share; O = Other statutes; WS = Water supplies

STATE	STATUTE	ENFORCEMENT AUTHORITIES	CLEANUP FUND	PRIORITY LIST	VICTIM COMPENSATION	CITIZEN SUIT PROVISION
Region IX						
Arizona	Environmental Quality Act	X	X (11)	X		X
California	Hazardous Substance Account Act	X	X	X	X (12)	X
Hawaii	Environmental Response Act	X	X	X		X
Nevada	Hazardous Waste Statute	X	ER, CS			X
Region X						
Alaska	Oil and Hazardous Substance Release Law	X	X	X		
	Hazardous Substance Release Control Law					
	Liability and Cost for Oil and Hazardous Substance Discharge Law	X				X
Idaho	Hazardous Waste Management Law		ER			
Oregon	Environmental Cleanup Law	X	X	X (13)		
Washington	Model Toxics Control Act	X	X			

ER = Emergency response and removal; CS = CERCLA share; O = Other statutes; WS = Water supplies

1. Limited to temporary resident relocation and temporary water supplies.
2. Reimbursement for costs of alternative water supplies or other emergency measures.
3. Additional funds provided by Environmental Quality Bond Act of 1986.
4. Enforcement authority limited to provision for joint and severe liability.
5. Creates repository for federal grant monies.
6. Establishes fund.: other statutes authorize fund uses.
7. Other fund established by Hazardous Substances Spill Prevention and Control Act.
8. Limited enforcement authority if groundwater affected.
9. Fund can also be used for related administrative costs.
10. Environmental Quality Restoration Fund, effective July 1, 1989.
11. Two additional funds, authorized under the Hazardous Waste Disposal Law, provide monies for limited cleanup activities.
12. Hazardous Substance Victims' Compensation Fund authorized, with appropriations up to $2 million per year. No appropriations to 1990.
13. Procedures for ranking sites under development.

Source: U.S. Environmental Protection Agency, *An Analysis of State Superfund Programs: 50-State Study,* Washington, DC: Office of Emergency and Remedial Response (1989), pp. 43–47.

1. Human beings and the environment are being exposed each year to a large number of chemical substances and mixtures.
2. Among the many chemical substances and mixtures which are constantly being developed and produced, there are some whose manufacture and use present an unreasonable risk of injury to health or the environment.
3. The effective regulation of interstate commerce of such chemical sub stances and mixtures also necessitates the regulation of intrastate commerce of such chemical substances and mixtures.

In order to solve the problems found, this act establishes these policies:

1. Adequate data should be developed with respect to the effect of chemical substances and mixtures on health and the environment and that the development of such data should be the responsibility of those who manufacture and distribute such chemical substances and mixtures.
2. Adequate authority should exist to regulate chemical substances and mixtures which present an unreasonable risk of injury to health or the environment, and to take action with respect to chemical substances and mixtures which are determined as imminent hazards.
3. Authority over chemical substances and mixtures should be exercised in such a manner as not to impede unduly or create unnecessary economic barriers to technological innovation while fulfilling the primary purpose of the Act to assure that such innovations and commerce in such chemical substances and mixtures do not present an unreasonable risk of injury to health or the environment.

The act recognized that there were insufficient data on many chemicals that had the potential to injure health or the environment. Consequently, the act established a testing procedure. As the test data become available, they are to be published in the *Federal Register.*

The act provides for the establishment of a committee to collect information such as the following:

1. The quantities of the substances or mixture that will be manufactured.
2. The quantities in which the substances or mixture enters the environment.
3. The number of individuals who are or will be exposed to the substances or mixtures in their places of employment and the duration of such exposure.
4. The extent to which the substances or mixtures are closely related to a chemical substance or mixture which is known to present an unreasonable risk of injury to health or the environment.
5. The effects of the substance or mixture on health or the environment.
6. The extent to which testing of the substances or mixtures may result in the development of data upon which the effects of the substance or

mixture on health or the environment can reasonably be determined or predicted.

7. The reasonably foreseeable availability of facilities and personnel for performing testing on the substances or mixtures.

The act requires that all manufacturers meet specified standards of health and environment. The manufacture of hazardous chemicals can be stopped by the administrator of the act if it presents a risk of injury to health or the environment.

For any person who violates the provisions of this act a civil penalty can be imposed in an amount not to exceed $25,000 for each violation. The administrator of the act on imposing a penalty takes into account the nature, circumstances, extent, and gravity of the violation.

A major feature of the act is that a citizen may present a petition to the administrator indicating that a particular chemical or substance presents a risk to health or the environment. In accordance with established procedures, the administrator may initiate action requested by the petitioners.

Solid Waste Disposal Act Amendment of 1980, Public Law 96-482
Amendments of 1984, Public Law 98-616

This act extends the 1965 act by requiring each state to develop an inventory of all sites at which hazardous waste has at any time been stored or disposed of. Such an inventory shall contain the following:

1. A description of the location of the sites at which any such storage or disposal has taken place before the date on which permits were required.
2. Such information will relate to the amount, nature, and toxicity of the hazardous waste at each site as may be practicable to obtain and as may be necessary to determine the health hazard associated with the site.
3. The name and address, or corporate headquarters of the owners of each site.
4. An identification of the types or techniques of waste treatment or disposal which have been used at each site.
5. Information concerning the current status of the site, including information as to whether hazardous waste is being treated or disposed of at such sites.

If a satisfactory program is not developed by the state, the Environmental Protection Agency shall carry out the inventory program.

If the EPA administrator determines that a site has hazardous waste, and if there is a release of this material presenting a hazard to human health or the environment, an order may be issued requiring the owner or operators to conduct monitoring, testing, and analysis resulting in a report on the conditions at the site.

Consumer Product Safety Amendments of 1981, Public Law 97-35
Title XII Omnibus Budget Reconciliation Act of 1981

Title XII of this act is concerned with consumer product safety. Its purpose is to identify hazardous products and the nature of the risk associated with these products.

A special feature of the act is the establishment of a Chronic Hazard Advisory Panel. A panel of seven members is appointed from a list of nominees of scientists prepared by the president of the National Academy of Sciences. The scientists on the panel have demonstrated their ability to assess chronic hazards and risks to human health presented by the exposure of humans to toxic substances or as demonstrated by exposure of animals to such substances. The panel must determine the risk of a product's being a carcinogen, mutogen, or a teratogen, and report its findings to the commission. In addition, the panel must attempt to estimate the probable harm to human health that will result from exposure to the substance.

Emergency Planning and Community Right-To-Know Act of 1986, Public Law 99-499
Superfund Amendments and Reauthorization Act of 1986, Title III

This act requires the governor of each state to appoint a State Emergency Response Commission, which is then required to appoint local emergency planning committees and then supervise and coordinate their activities. In addition, the commission is to designate emergency planning districts in order to facilitate preparation and implementation of emergency plans. The plans may involve more than one state. Each of the local committees shall include representatives from each of the following groups—state and local officials; law enforcement, civil defense, fire fighting, first aid, health, local environmental, hospital, and transportation personnel; broadcast and print media; community groups; and owners and operators of facilities subject to the requirements of the act. The local emergency planning groups shall establish procedures for receiving and processing public requests.

A major stipulation of the act is the development of a Comprehensive Emergency Response Plan, which is to include the following:

1. Resources needed to develop and implement an emergency plan.
2. Identification of facilities and routes likely to be used for the transportation of hazardous substances.
3. Methods and procedures to be followed by facility owners and operators and local emergency and medical personnel to respond to any release of hazardous or toxic substances.
4. Designation of a community emergency coordinator and facility emergency coordinators, who shall make determinations necessary to implement the plan.
5. Procedures providing reliable, effective, and timely notification by the facility emergency coordinator and the community emergency coordinator to persons designated in the emergency plan and to the public that a hazardous release has occurred.
6. Methods of determining the occurrence of a release, and the area of population likely to be affected by such release.
7. A description of emergency equipment and facilities in the community and at each facility in the community, and identification of the persons responsible for such equipment and facilities.
8. Evacuation plans, including provisions for a precautionary evacuation and alternative traffic routes.
9. Training programs, including schedules for training of local emergency response and medical personnel.
10. Methods and schedules for exercising the emergency plan.

The act requires an owner or operator of any facility producing hazardous chemicals to prepare a material safety data sheet. It also requires that emergency and hazardous chemical inventory forms be prepared. The toxic chemicals subject to these requirements include the following:

1. A chemical that is known to cause or can reasonably be anticipated to cause significant adverse acute human health effects at concentration levels that are reasonably likely to exist beyond facility site boundaries as a result of continuous, or frequently recurring releases.
2. A chemical that is known to cause or can reasonably be anticipated to cause in humans:
   a. Carcinogen or teratogenic effects.
   b. Serious or irreversible reproductive dysfunctions, neurological disorders, heritable genetic mutations or other chronic health effects.
3. A chemical that is known to cause, or can reasonably be anticipated to cause, because of its toxicity, its toxicity and persistence in the environment, or its toxicity and tendency to bioaccumulate in the environment, a significant effect on the environment of sufficient seriousness to warrant reporting.

Pipeline Safety Act of 1979, Public Law 96-129 Amendment of the National Gas Pipeline Safety Act of 1968 and Title II, Hazardous Liquid Pipeline Safety Act of 1979

This act establishes minimum safety standards for the transportation of hazardous liquids and gases. The standards apply to each person who engages in the transportation of hazardous liquids or who owns or operates pipeline facilities. The standards may apply to the design, installation, inspection, emergency plans and procedures, testing, construction, extension, operation, replacement, and maintenance of pipeline facilities.

The secretary of transportation may find a pipeline facility hazardous for the following reasons:

> If the pipeline facility has been constructed or operated with any equipment, material, or technique which is hazardous to life or property. This involves the characteristics of the pipe and other equipment, the nature of the materials transported, and aspects of the areas in which the pipeline facility is located, in particular the climatic and geologic conditions.

Pesticide Legislation

Federal Insecticide, Fungicide and Rodenticide Act of 1947, Public Law 1033

This act was enacted to regulate the marketing of economic poisons and devices. The term *economic poison* referred to any substance or mixture of substances intended for preventing, destroying, repelling, or mitigating any insects, rodents, fungi, weeds, and other forms of plant or animal life or viruses. The term *insecticide* meant any substance intended for preventing, destroying, repelling, or mitigating any insect that may be present in the environment.

The act had the following stipulation for use of economic poisons. The substance was considered misbranded:

1. If it was an imitation of or was offered for sale under the name of another economic poison.
2. If the labeling accompanying it did not contain a warning or caution statement which may be necessary to prevent injury to living man, animals, vegetation, and invertebrate animals.

3. If the label did not bear an ingredient statement on the container.
4. If in the case of an insecticide, fungicide, or herbicide, when used as directed or in accordance with commonly recognized practice, it shall be injurious to living man, or other animals or vegetation, except weeds, to which it was applied, or to the person applying economic poison.

The act prohibits the following:

The selling of any economic poison unless it was properly labelled. The poisons commonly included were lead arsenate, calcium arsenate, magnesium arsenate, zinc arsenate, zinc arsenite, sodium fluoride, sodium fluosilicate, and barium fluosilicate.

This early act did not control the sale of economic poisons. It required only that proper labeling be adhered to.

Federal Environmental Pesticide Control Act of 1972, Public Law 95-516
Amends the Federal Insecticide, Fungicide and Rodenticide Act of 1947, Amended 1978 and 1980

This act completely rewrites the initial act of 1947. It strengthens many sections of the original act in order to tighten control of pesticide registration, to ensure protection of humans and the environment. The new bill has the following features:

1. Requires that no person can distribute or sell a pesticide which is not registered.
2. Restricts the use of pesticides to persons who are certified competent to handle the material.
3. Extends the registration to all pesticides whether it is used within the state or enters into interstate commerce.
4. Prohibits the use of any pesticide in a manner inconsistent with its labeling. No pesticide could be registered or sold unless its labeling was such as to prevent any injury to man or any substantial adverse effect on environmental values, taking into account the public interest, including benefits from it use.
5. Requires pesticides to be classified for general or restricted use. Restricted use of pesticides could be used only by or under the supervision of certified applicators.
6. Strengthens enforcement by requiring the registration of all pesticides producing establishments, authorizing the entry of establishments and other places where pesticides are held for sale or distribution, authorizing the stoppage of sales, use or removal orders, authorizing the cooperation with states and improving procedures governing registration.

7. Gives the manufacturer proprietary rights in their test data.
8. Establishes pesticide packaging standards to protect children and adults.
9. Provides for certification of applicators by the states.

The stricter regulation policies will aid in determining violations and accidental use of pesticides. A sale of a pesticide may be stopped if violations in registration are detected in dangerous pesticides. These improved procedures will remove undesirable pesticides from the market and speed development and use of desirable pesticides. Giving manufacturers the proprietary rights to test data will encourage applicants to assume their safety.

Pesticide Monitoring Improvements Act of 1988, Public Law 100-418
Omnibus Trade and Competitiveness Act of 1988, Agricultural Trade, Subtitle G

This act furthers the control of pesticides in order to assure public health. It establishes a data management system for the following purposes:

1. Record, summarize, and evaluate the results of the program for monitoring food products for pesticide residues.
2. Identify gaps in the pesticide monitoring program including food products and food from foreign countries as well as from domestic sources.
3. Detect trends in the presence of pesticide residues in food products and identify public health problems emerging from the occurrence of pesticide residues in food products.
4. Focus testing resources for monitoring pesticide residues in food in order to detect those residues which pose a public health concern.
5. Prepare summaries of information.

The Food and Drug Administration has the responsibility of developing a computerized data management system. The information is to include the following:

1. The types of imported and domestically produced food products analyzed for compliance with the requirements of the Federal Food, Drug, and Cosmetic Act regarding the presence of pesticide residues.
2. The pesticide residues which may be detected using the testing methods employed.
3. The pesticide residues in such food detected and the levels detected.
4. The compliance status of each sample of such food tested and the violation rate for each country-product combination.

5. The action taken with respect to each sample of such food found to be in violation of the Federal Food, Drug and Cosmetic Act and its ultimate disposition.

The Department of Health and Human Services has the responsibility to develop cooperation agreements with the governments of other countries that are the major sources of food imports into the United States subject to pesticide residue monitoring. The agreements are to provide information identifying each of the pesticides used in the production, transportation, and storage of food products imported from production regions of such countries.

In addition, the secretary of the Department of Health and Human Services, in consultation with the administrator of the Environmental Protection Agency, shall:

1. Develop a detailed long-range plan and timetable for research for the development of and validation of new and improved analytical methods capable of detecting at one time the presence of multiple pesticide residues in foods.
2. Conduct a review to improve the cost-effectiveness of monitoring and enforcement activities under the Federal Food, Drug and Cosmetic Act.

Hazardous Mineral Legislation

Asbestos School Hazard Detection and Control Act of 1980, Public Law 96-270

This act was passed in response to the following findings of Congress:

1. Exposure to asbestos fibers has been identified over a long period of time and by reputable medical and scientific evidence as significantly increasing the incidence of cancer and other severe or fatal diseases, such as asbestosis.
2. Medical evidence has suggested that children may be particularly vulnerable to environmentally induced cancers.
3. Medical science has not established any minimum level of exposure to asbestos fibers which is considered safe to individuals exposed to the fibers.
4. Substantial amounts of asbestos, particularly in sprayed form have been used in school buildings, especially during the period 1946 to 1972.
5. Partial surveys in some states indicated that (a) in a number of school buildings materials containing asbestos fibers have become damaged or friable, causing asbestos fibers to be dislodged into the air, and (b) asbestos concentrations far exceeding normal ambient air levels have been found in school buildings containing such damaged materials.

6. The presence in school buildings of friable or easily damaged asbestos creates an unwarranted hazard to the health of the school children and school employees who are exposed to such materials.

7. The Department of Health and Human Services and the Environmental Protection Agency, as well as several states have attempted to publicize the potential hazards to school children and employees from exposure to asbestos fibers, but there is no systematic program for identifying hazardous conditions in schools or for remedying those conditions.

8. Because there is no Federal health standard regulating the concentration of asbestos fibers in noncommercial workplace environments such as schools, school employees and students may be exposed to hazardous concentrations of asbestos fibers in the school buildings which they use each day.

9. Without an improved program of information distribution, technical and scientific assistance, and financial support, many local educational agencies and states will not be able to mitigate the potential asbestos hazard in schools.

10. The effective regulation of interstate commerce for the protection of the public health requires the establishment of programs under this Act to identify and mitigate hazards from exposure to asbestos fibers and materials emitting such fibers.

The purposes of this act were as follows:

1. Direct the Secretary of Education to establish a task force to assist states and local educational agencies to ascertain the extent of the danger to the health of school children and employees from asbestos materials in schools.

2. Requires states receiving administrative funds for any applicable program to prepare a plan describing the manner in which information relating to programs established under this Act shall be distributed to local educational agencies.

3. Provides scientific, technical, and financial assistance to state educational agencies and local educational agencies to enable them to conduct an asbestos detection program to identify asbestos hazards in schools.

4. Provides loans to local educational agencies for the mitigation of asbestos hazards which constitute an imminent hazard to the health and safety of school children and employees.

5. Assure that no employee of any local educational agency suffers any disciplinary action as a result of calling attention to potential asbestos hazards which may exist in schools.

If a state accepted administrative funds for development of a program, the state was required to submit a plan with these features:

1. Describes the manner in which the state is to distribute information on asbestos to the local educational agencies.

2. Contains a general description of the contents of the information.
3. Describes procedures to be used by the state for monitoring records.
4. Designates a state agency to administer the program.

The duties of the task force were as follows:

1. The compilation of medical, scientific, and technical information.
2. The distribution of the information to the states.
3. The review of the grants and loans.
4. The review of guidelines established by the Environmental Protection Agency for identifying those schools in which exposure to asbestos fibers constitutes a health problem and for taking appropriate corrective actions at such schools.
5. Providing the Secretary of Education with assistance in formulating standards and procedures.

Asbestos Hazard Emergency Response Act of 1986, Public Law 99-519

This act was enacted to provide regulatory guidance for the removal of asbestos from schools. In addition, there was no uniform program for accrediting persons involved in asbestos identification and abatement, and local educational agencies were not required to use accredited contractors for asbestos works. Because federal standards were lacking, the control of asbestos was difficult.

The purpose of this act was to remedy some of these problems:

1. To provide for the establishment of Federal regulations which require inspection for asbestos-containing material and implementation of appropriate response actions with respect to asbestos-containing material in the nation's schools in a safe and complete manner.
2. To mandate safe and complete periodic reinspection of school buildings following response actions, where appropriate.
3. To require the Environmental Protection Agency to conduct a study to find out the extent of the danger to human health posed by asbestos in public and commercial buildings and the means to respond to any such danger.

Within 360 days of the passage of this legislation, the Environmental Protection Agency was required to develop regulations. A major feature of these regulations was the inspection of asbestos-containing schools and the response actions required depending upon the condition of the asbestos. The act also required that standards be established for the transportation and disposal of the asbestos. Lack of enforcement of the act resulted in fines of not more than $5,000. Any person could petition the EPA to initiate a petition to enforce the law, and any person could commence a civil suit against the EPA to compel action by the agency to comply with the law.

School Asbestos Management Plans Act of 1988, Public Law 100-368

This act amends the provisions of the Toxic Substances Control Act relating to asbestos in the nation's schools by providing adequate time for local educational agencies to submit asbestos management plans to state governors and to begin implementation of these plans. In order to develop satisfactory plans a local agency may request deferral. The Environmental Protection Agency was to list in the *Federal Register* approved asbestos removal training courses. A major feature of the act was worker protection. It specifically states that a local educational agency may not:

> Perform, or direct an employee to perform, renovations or removal of building materials, except emergency repairs, in the school, unless (a) the school is carrying out work under a grant awarded under the Asbestos School Hazard Abatement Act of 1984, or (b) an inspection that complies with the Code of Federal Regulations.

Any employee who is directed to conduct emergency repair involving any building material containing asbestos shall:

1. Be provided the proper training to safely conduct such work in order to prevent potential exposure to asbestos.
2. Be provided the proper equipment and allowed to follow work practices that are necessary to safely conduct such work in order to prevent potential exposure to asbestos.

Asbestos Information Act of 1988, Public Law 100-577

This act requires that any person who manufactured or processed, before October 31, 1988, asbestos or asbestos-containing materials, thermal system insulation, or miscellaneous material in building shall submit to the administrator of the Environmental Protection Agency the years of manufacture, the types or classes of products, and, to the extent available, other identifying characteristics reasonably necessary to identify or distinguish the asbestos or asbestos-containing materials. The EPA administrator will then take the following steps:

1. Review the information submitted by the manufacturer.
2. Analyze such information to determine whether it is reasonably necessary to identify or distinguish the particular asbestos or asbestos-containing material.

Lead Contamination Control Act of 1988, Public Law 100-572

This act amends the Public Health Services Act, Title XIV, Safe Drinking Water Act, to control lead in drinking water. The standard for "lead free" with respect to drinking water coolers is that each part or component of the cooler that may be in contact with drinking water contain not more than 8 percent lead, or, the container cannot contain any solder, flux or storage tank interior with more than 0.2 percent lead. All such containers must be recalled from use.

In addition, this act amends the Public Health Service Act in order to discover and prevent lead poisoning. Grants may be made to local governments for the initiation and expansion of community programs designed for the following purposes:

1. Screen infants and children for elevated blood lead levels.
2. Assume referral for treatment of, and environment intervention for, infants and children with such blood levels.
3. Provide education about childhood lead poisoning.

In order to control lead poisoning, grants are made to local governments. The grant proposals must include the following:

1. A complete description of the program.
2. Educational programs designed to communicate to parents, educators, and local health officials the significance and prevalence of lead poisoning in infants and children which the program is designed to detect and prevent.
3. A quantity report on number of infants and children screened for elevated lead blood levels and type of medical referrals.

Act to Prevent Pollution from Ships, Public Law 96-478

This is an act to implement the Protocol of 1978, relating to the International Convention for the Prevention of Pollution from Ships, 1973, known as MARPOL Protocol. By this act the Coast Guard establishes standards applicable to ships entering American ports. Standards are also established determining the adequacy of port facilities. Penalties are imposed if standards are not met.

4

Directory of Organizations

THE ORGANIZATIONS THAT CONSIDER the problems of hazardous material and toxic waste can be grouped into three major categories— U.S. government agencies (including government advisory committees), private organizations, and international organizations. Of the U.S. government agencies, the Environmental Protection Agency plays a prominent role in establishing policies and enforcing regulations. There are many intergovernmental committees that have specific and limited objectives. Many provide information to the EPA. These agencies may be permanent organizations, or they may be abolished when their objectives are achieved. The private organizations usually have been developed to solve specific problems, such as transportation of hazardous material, cleanup of abandoned waste sites, or management of solid wastes. Most of these have been formed since 1980 as the public has become aware of the problems of hazardous material and toxic waste. The international organizations developed as it became evident that the problems of hazardous and toxic wastes were worldwide in scope.

U.S. Government Agencies

Environmental Protection Agency
401 M Street, SW
Washington, DC 20460

Description: The Environmental Protection Agency protects and enhances our environment today and for future generations to the fullest extent possible under the laws enacted by Congress. The EPA was established as an independent agency in the executive branch in 1970.

Purpose: The EPA's mission is to control and abate pollution in the areas of air, water, solid waste, pesticides, radiation, and toxic substances. Its mandate is to mount an integrated, coordinated attack on environmental pollution in cooperation with state and local governments.

Activities: The agency endeavors to abate and control pollution systematically, by proper integration of a variety of research, monitoring, standard-setting, and enforcement activities. In addition, the EPA coordinates and supports research and antipollution activities by state and local governments, private and public groups, individuals, and educational institutions. It also reinforces efforts among other federal agencies with respect to the impact of their operations on the environment. It is required to publish its findings when it finds an infraction in regard to public health, welfare, or environmental quality. The agency is designed to serve as the public's advocate for a livable environment.

Office of Solid Waste and Emergency Response

This office provides policy, guidance, and direction for the EPA's hazardous waste and emergency response programs. Its functions include the following:

1. Development of policies, standards and regulations for hazardous waste treatment, storage and disposal
2. National management of the Superfund toxic waste cleanup program
3. Development of guidelines for the emergency preparedness and "Community Right To Know" programs
4. Development of guidelines and standards for underground storage tanks
5. Enforcement of applicable laws and regulations
6. Analysis of technologies and methods for the recovery of useful energy from solid waste
7. Provision of technical assistance in the development, management, and operation of waste management activities

Office of Pesticides and Toxic Substances

This office is responsible for the following:

1. Developing national strategies for the control of toxic substances
2. Directing pesticides and toxic substances enforcement activities

3. Developing criteria for assessing chemical substances, standards for test protocols for chemicals, rules and procedures for industry reporting, and regulations for the control of substances deemed to be hazardous to humans or the environment
4. Evaluating and assessing the impact of existing chemicals, new chemicals, and chemicals with new uses to determine hazards and, if needed, to develop appropriate restrictions

Additional activities are aimed at controlling and regulating pesticides and reducing their use to ensure human safety and protection of environmental quality. These include the following:

1. Establishment of tolerance levels for pesticides that occur in or on food
2. Monitoring of pesticide residue levels in food, humans, and non-target fish and wildlife and their environments
3. Investigation of pesticide accidents
4. Coordination of activities, under the office's statutory responsibilities, with other agencies for the assessment and control of toxic substances and pesticides

Office of Research and Development

This office is responsible for a national research program seeking ways to have technological control over all forms of pollution. It directly supervises the research activities of the EPA's national laboratories and gives technical policy direction to those laboratories that support the program responsibilities of the required offices. Close coordination of the various research programs is designed to yield a synthesis of knowledge from the biological, physical, and social sciences that can be interpreted in terms of human and environmental needs.

Other offices of the EPA include the Office of Air and Radiation and the Office of Water.

Regional Offices: The EPA has ten regional offices. Their basic function is to develop strong local programs for pollution abatement. The regional administrator is the agency's principal representative in each region in contacts and relationships with federal, state, interstate, and local agencies, industry, academic institutions, and other public and private groups. The locations of the ten regional offices are as follows:

1. Boston, MA 02203
   John F. Kennedy Federal Building
2. New York, NY 10278
   26 Federal Plaza
3. Philadelphia, PA 19107
   841 Chestnut Street

    4. Atlanta, GA 30365
       345 Courtland Street, NE
    5. Chicago, IL 60604
       230 South Dearborn Street
    6. Dallas, TX 75270
       1201 Elm Street
    7. Kansas City, KS 66101
       726 Minnesota Street
    8. Denver, CO 80202
       999 18th Street
    9. San Francisco, CA 94105
       215 Fremont Street
   10. Seattle, WA 98101
       1200 Sixth Avenue

Agency for Toxic Substances and Disease Registry
200 Independent Avenue, SW
Washington, DC 20201

Description: This is an operating agency established within the Public Health Service by the Department of Health and Human Services on April 19, 1983. The agency's mission is to carry out the health-related responsibilities of the Comprehensive Environmental Response, Compensation and Liability Act of 1980, as amended by the Superfund Amendments and Reauthorization Act of 1986, the Resources Conservation and Recovery Act, and provisions of the Solid Waste Disposal Act relating to sites and substances found at those sites and other forms of uncontrolled releases of toxic substances into the environment.

Purpose: The agency provides leadership and direction to programs and activities designed to protect both the public and workers from exposure and/or the adverse health effects of hazardous substances in storage sites as released in fires, explosions, or transportation accidents.

Activities: The agency, in cooperation with states and other federal and local agencies, collects, maintains, analyzes, and disseminates information relating to serious diseases, mortality, and human exposure to toxic or hazardous substances; establishes appropriate registries necessary for long-term follow-up or specific scientific studies; establishes and maintains a complete list of areas closed to the public or otherwise restricted in use because of toxic substance contamination; assists, consults, and coordinates with private and public health care providers in the provision of medical care and testing of exposed individuals; assists the Environmental Protection Agency in identifying hazardous waste substances to be regulated; develops scientific and technical procedures for evaluating public health risks from hazardous substance incidents and for developing recommendations to protect public health and worker safety and health in instances of exposure or potential exposure to hazardous substances; and arranges for program support to ensure adequate response to public health emergencies.

Publications: None

Department of Justice
Land and Natural Resources Division
Constitution Avenue and Tenth Street, NW
Washington, DC 20530

Description: The Land and Natural Resources Division of the Department of Justice represents the United States in litigation involving public lands and natural resources and environmental quality.

Activities: The fastest-growing area of responsibility for this division involves civil and criminal enforcement of environmental statutes. The division brings civil and criminal enforcement cases primarily on behalf of the Environmental Protection Agency for the control and abatement of pollution of air and water resources, the regulation and control of toxic substances, and the environmental hazards posed by hazardous wastes. In the hazardous waste area, most cases are brought under the Comprehensive Environmental Response, Compensation and Liability Act of 1980, as amended by the Superfund Amendments and Reauthorization Act of 1986, for the purpose of protecting public health and ensuring that responsible parties, rather than the public, bear the burden of abating hazardous waste pollution.

The various environmental enforcement statutes also generate a great deal of defensive litigation, since many of the provisions of those laws also apply to the federal programs or activities. This litigation generally falls into one of two categories: suits by industry and environmental groups challenging agency regulations, and challenges to agency decisions relating to individual activities such as permits, grants, and variances.

Publications: None

Office of Hazardous Materials Transportation
400 Seventh Street, SW
Washington, DC 20590

Description: This office develops and issues regulations for the safe transportation of hazardous materials by all modes, excluding bulk transportation by water. The regulations cover shipping and carrier operations, packaging and container specifications, and hazardous materials definitions. The office is also responsible for the enforcement of regulations other than those applicable to a single mode of transportation.

Activities: The office reviews and analyzes reports made by industry and by field staff bearing upon compliance with the regulations, and offers a number of training and information dissemination activities designed to familiarize industry personnel with the requirements of hazardous material regulations, to educate federal and state inspectors in enforcement procedures, and to act as a national focal point for coordination and control of the multimodal hazardous material regulatory program, ensuring uniformity of approach and action.

Publications: None

U.S. Intergovernmental Advisory Committees

Advisory Committee To Negotiate Regulation Governing Major and Minor Modifications of Resources (Hazardous Wastes)
Regulatory Negotiation Project
Environmental Protection Agency
401 M Street, SW
Washington, DC 20460

Description: This committee was established in 1986 as a public advisory group to the Environmental Protection Agency. It consists of no more than 25 members interested in waste management.

Activities: The committee makes recommendations to amend current regulations governing major and minor modifications to Resource Conservation and Recovery Act permits for hazardous waste management facilities.

Publications: None

Advisory Council on Hazardous Substances Research and Training (Hazardous Materials)
National Institutes of Health
9000 Rockville Pike, Building 1
Bethesda, MD 20892

Description: The council was established March 9, 1987, by the secretary of the Department of Health and Human Services.

Purpose: This body serves as a public advisory council of the National Institutes of Environmental Health Sciences, National Institutes of Health, Public Health Services, Department of Health and Human Services.

Activities: The council advises the agencies mentioned regarding the hazardous substances research and training program. The program is mandated to be established and supported by the secretary of health and human services under Section 311 (a) (5) of P.L. 96-510, the Comprehensive Environmental Response, Compensation and Liability Act, as amended by P.L. 99-499, the Superfund Amendments and Reauthorization Act of 1986.

Publications: None

Interstate Low-Level Radioactive Waste Commissions

Central Interstate Low-Level Radioactive Waste Commission
Office of Air Quality and Nuclear Energy
Department of Environmental Quality
P.O. Box 14690
Baton Rouge, LA 70898

Midwest Interstate Low-Level Radioactive Waste Commission
350 North Robert Street, Room 558
St. Paul, MN 55101

Northeast Interstate Low-Level Radioactive Waste Commission
55 Princeton Hightstown Road
Princeton Junction, NJ 08550

Northwest Low-Level Waste Compact Committee
Low-Level Radioactive Waste Program
Office of Nuclear Waste
Washington Department of Ecology
Mail Stop PV-11
Olympia, WA 98504

Rocky Mountain Low-Level Radioactive Waste Board
1600 Stout Street, Suite 100
Denver, CO 80202

**Southeast Interstate Low-Level Radioactive Waste
Management Commission**
3901 Barrett Drive, Suite 100-B
Raleigh, NC 27609

Description: The regional commissions were established to enter into agreement with any person, state, or group of states for the right to use regional facilities for low-level radioactive waste generated outside each region and to use waste disposal facilities outside the designated region. They are interstate regulatory commissions.

Activities: The commission will appear as an intervenor or party in interest before any court of law or federal, state, or local agency, board, or commission on any matter related to low-level radioactive waste management; reviews the emergency closure of a regional facility, determines the appropriateness of the closure, and takes necessary action to ensure that the interests of the region are protected. Each commission provides for sufficient facilities for the proper management of low-level radioactive waste generated in the region, provides for the protection of the health and safety of residents of the region, establishes the number of facilities required to manage low-level radioactive waste management in the region, ensures the ecological and economical management of low-level radioactive waste, and works for the promotion of above-ground facilities and other disposal technologies for greater and safer confinement of low-level radioactive waste than shallow burial facilities.

Hazardous Waste/Superfund Research Committee
Office of Solid Waste Management and Emergency Response
Office of Research and Development, R.D. 674
Environmental Protection Agency
401 M Street, SW
Washington, DC 20460

Description: The committee consists of managers from the Office of Research and Development, EPA.

Activities: The committee develops research needs, suggests research programs to meet those needs, and advises the Office of Research and Development on matters regarding hazardous wastes.

Publications: None

Interagency Ocean Disposal Program Coordinating Committee (Waste Disposal)
Office of Water and Hazardous Materials
Oil and Special Materials Control Division
Environmental Protection Agency
401 M Street, SW
Washington, DC 20460

Description: This committee functions under the Marine Protection Research and Sanctuaries Act of 1972, which regulates dumping of all types of materials into ocean waters. The committee coordinates the work of allied agencies such as National Oceanic and Atmospheric Administration, Corps of Engineers, and Council of Environmental Quality.

Activities: Responsibilities of the committee are to develop an integrated approach toward all aspects of implementing the ocean dumping legislation and to coordinate the activities of governmental agencies with regard to ocean dumping. Protocols are being established for disposal site evaluations, monitoring, and research studies to guide the field efforts and data interpretation procedures of all agencies involved.

Publications: None

Interagency Toxic Substances Data Committee (ITSDC)
Office of Program Integration and Information
Office of Toxic Substances
Environmental Protection Agency
401 M Street, SW
Washington, DC 20460

Description: This committee was established under the authority of P.L. 94-469, the Toxic Substances Control Act, P.L. 91-190, National Policy Act of 1969, and P.L. 91-224, the Environmental Quality Improvement Act of 1970. It consists of 14 representatives of various federal government departments and agencies.

Activities: The committee coordinates the planning and activities on chemical data and information of the leading federal producers and users of chemical data in order to minimize the reporting problems faced by the private sector. It also aids in the collection, analysis, and exchange of data among federal agencies and other groups.

Publications: None

National Response Team (Hazardous Materials)
Environmental Protection Agency
401 M Street, SW
Washington, DC 20460

Description: Established on January 23, 1987, by Presidential Executive Order. The committee consists of governmental agencies.

Activities: This is an interagency advisory group established to plan and coordinate preparedness and response actions for release or potential release of hazardous substances.

Publications: None

TSCA Interagency Testing Committee (Toxic Substances)
401 M Street, SW
Washington, DC 20460

Description: This committee is a group of agencies such as the Council on Environmental Quality, National Institute of Environmental Health Sciences, Departments of Agriculture, Defense, and Interior, and 11 others that function under the Environmental Protection Agency.

Activities: The committee recommends chemical substances and mixtures to which the Environmental Protection Agency should give priority consideration for the promulgation of testing rules.

Publications: Annual report to the EPA, in the *Federal Register,* revising and updating the committee's priority list of chemicals

Wastes in Marine Environments Advisory Panel (Waste Disposal)
Oceans and Environment Program
Science, Information and Natural Resources Division
Office of Technology Assessment
U.S. Congress
Washington, DC 20510

Description: This is a public advisory panel established in 1984, on the recommendation of Congress, that functions under the Ocean and Environment Program, Science, Information and Natural Resources Division and the Office of Technology Assessment.

Activities: The panel reviews and advises on the study plans, work in progress, and reports of the Office of Technology Assessment's study of wastes in the marine environment. Stress is placed on the study of whether or not the oceans can be used for disposal, what specific wastes can be disposed of in marine environments, and what technologies can be used.

Publications: Subseabed Disposal of High-Level Radioactive Waste Ocean Incineration: Its Role in Managing Hazardous Wastes in Marine Environments: Their Management and Disposal; occasional reports

Private Organizations in the United States

Association of State and Territorial Solid Waste Management Officials (ASTSWMO)
444 North Capitol Street, NW, Suite 388
Washington, DC 20001

Description: Founded in 1974, the association has 250 members and staff of 7. It represents state solid and hazardous waste directors.

Purpose: The association was formed to develop programs in the management of solid wastes by promoting enforcement of all pertinent laws and federal regulations.

Activities: The association furnishes information on technology and management techniques, studies and analyzes critical issues, trains state employees in areas of waste management, conducts solid waste management training workshops, and compiles statistics on waste management. The association holds semiannual conferences.

Publications: State Waste Management Program Officials (annual); *Directory; Washington Update* (10/year); *Newsletter; Information Brochure* (annual); surveys

Center for Hazardous Materials Research (CHMR)
320 William Pitt Way
University of Pittsburgh Applied Research Center
Pittsburgh, PA 15238

Description: This nonmembership center was founded in 1985 and has a staff of 25.

Purpose: The goal of the center is to develop practical solutions to problems associated with hazardous waste management.

Activities: The center develops and implements policy on the economic, environmental, technological, public health, and institutional issues caused by hazardous materials and wastes. It works with industry and government agencies, and offers a student internship program as well as educational and training programs in hazardous materials, health and safety, and pesticide certification. The center also provides technical assistance to industry for pesticide registration. A speakers' bureau is available. The center's library has 500 volumes, and its database contains technical, regulatory, and vendor information. The center also operates a telephone hot line.

Publications: The Minimizer (quarterly); *Newsletter*

Citizen's Clearinghouse for Hazardous Wastes (Environment) (CCHW)
P.O. Box 926
Arlington, VA 22216

Description: The CCHW was founded in 1981. It currently has a membership of 13,000, with a budget of $500,000. Involved are 50 regional groups, 250 state groups, and 3,560 local groups. Members include people who live near hazardous waste dumps.

Purpose: This group is concerned with the possibility of adverse health effects on adults and children from contact with toxic chemicals and hazardous wastes. It was established as a result of the many problems faced by the Love Canal Homeowners Association.

Activities: CCHW representatives visit sites to determine how severe their problems are, and furnish information and guidance in dealing with toxic waste problems. The organization asks members to urge their lawmakers to take action. It also conducts research to decide the dangerous levels of usage of chemicals. The group maintains a speakers' bureau and a 2,000-volume library on waste management technology, and presents programs on disposal technologies and public policy issues. The CCHW also has state and regional leadership training programs. The organization holds triennial conferences.

Publications: Action Bulletin (4/year); *Everyone's Backyard* (4/year); more than 50 additional titles

Clean Sites, Inc. (Environment) (CSI)
1199 North Fairfax Street
Alexandria, VA 22314

Description: This nonmembership organization was founded in 1984 and has a budget of $4.5 million. It consists of representatives of environmental organizations and industries.

Purpose: CSI brings together those responsible for abandoned waste sites and negotiates to determine who will be responsible for the cleanup.

Activities: The organization selects hazardous waste sites that need attention and arranges for cleanup. It provides financial management services to collect, invest, and disburse responsible parties' funds for cleanup. The group also conducts seminars on cleanup of hazardous waste. CSI consists of three sections: Project Management and Public Policy and Education Division, Settlement Division, and Technical Division. It holds six seminars a year.

Publications: Annual Report; *Clean Sites Forum Newsletter* (quarterly); *Allocation of Superfund Site Costs through Mediation*; brochures

Concerned Neighbors in Action (CNA)
P.O. Box 3847
Riverside, CA 92519

Description: This group was founded in 1979 and has a membership of 2,000. It is composed of families living near the Stringfellow acid pits, a hazardous waste dump near Los Angeles.

Purpose: The group was formed to advocate the cleanup of the Stringfellow site.

Activities: CNA distributes information on hazardous waste throughout the nation even though it is based in California. The group also compiles statistics and maintains a speakers' bureau. It holds monthly meetings.

Publications: CNA Newsletter (quarterly); *Membership Directory* (periodic)

Governmental Refuse Collection and Disposal Association (GRCDA)
8750 Georgia Avenue, Suite 123
Silver Spring, MD 20910

Description: This organization, founded in 1961, has 3,000 members and a staff of 9, with a budget of $1 million. It has 25 state groups, and membership includes public agency officials, private corporate officials such as employees, managers of solid waste management agencies, and consultants.

Purpose: GRCDA's goal is to improve solid waste management services through training.

Activities: The organization conducts 15–25 seminars annually and also sponsors training programs. It maintains a library of 6,000 documents on solid and hazardous waste management. Committees within GRCDA deal with such areas as collection, disposal, hazardous waste, and resource recovery. The group holds an annual conference that is coupled with an equipment show.

Publications: Newsletter (monthly); *Proceedings of Meetings*

Hazardous Materials Advisory Council
1012 14th Street, NW, Suite 907
Washington, DC 20005

Description: Founded in 1978, this council has 230 members and a staff of 6. Membership is made up of shippers, carriers, and container manufacturers of hazardous materials and waste shipper and carrier associations.

Purpose: The council works to promote safe transportation of hazardous materials, substances, and wastes, and provides assistance in answering regulatory questions, guidance to appropriate governmental resources, and advice in establishing corporate compliance and safety programs.

Activities: The council holds seminars on hazardous domestic and international materials packaging and transporting, and presents the George L. Wilson Memorial Award for outstanding achievements in hazardous materials transportation and safety by a person or company. It holds an annual convention/ meeting. The council was formerly known as the Hazardous Materials Advisory Committee (1978).

Publications: Courier (monthly); *Directory* (annual)

Hazardous Materials Control Research Institute (HMCRI)
9300 Columbia Boulevard
Silver Spring, MD 20910

Description: Founded in 1976, HMCRI has a membership of 3,500 and a staff of 12. The institute is affiliated with the American Society of Civil Engineers and the National Solid Waste Management Association. Membership consists of corporations, engineers, consultants, professors, scientists, government and corporate administrators, and others concerned with the safe management of hazardous materials and waste prevention, cleanup, and control.

Purpose: This organization's goals are to promote the use of risk assessment methods to achieve a balance between continuing industrial growth and a livable environment and to help minimize the hazardous materials discharged or released into the environment. HMCRI favors reasonable standards, monitoring, test procedures, and reporting requirements.

Activities: HMCRI disseminates information about technical advancements and institutional requirements in the disposal of hazardous waste, holds training programs on the control and management of toxic and hazardous materials, and gives information about the control of hazardous materials to industry, academia, the public, and regulatory agencies. The institute also maintains a 500-volume library on environmental topics. The group holds semiannual conferences.

Publications: Focus (monthly); *Newsletter; Hazardous Materials Control Directory* (biennial); *Hazardous Materials Control Magazine* (bimonthly); *Journal on Hazardous Waste and Hazardous Materials* (quarterly); *Management of Hazardous Wastes and Hazardous Materials* (annual); *Management of Uncontrolled Hazardous Waste Sites* (annual)

Hazardous Materials Systems (Bureau of Explosives) (HMS)(BOE)
50 F Street, NW
Washington, DC 20001

Description: This organization, known until 1986 as the Bureau of Explosives, was founded in 1907. It has a membership of 600 and a staff of 38, and a $2.4 million budget. Railroads, steamship companies, motor carriers, manufacturers, shippers of hazardous materials, and container manufacturers are members.

Purpose: The organization provides emergency procedures in the handling of hazardous articles.

Activities: The group conducts research and educational programs in safe transportation and storage of explosives and other hazardous materials, and provides ICARIS, an online hazardous materials database.

Publications: Tariff (annual) Department of Transportation Regulations; *Emergency Handling of Hazardous Materials in Surface Transportation*; *Emergency Action Guide*; posters

Hazardous Waste Federation (HWF)
c/o New Mexico Hazardous Waste Management Society
Division 3314
Albuquerque, NM 87185

Description: Founded in 1986, the federation is composed of eight local groups, including local and regional associations made up of people engaged in the management of hazardous wastes and materials.

Purpose: HWF's goals are to increase public awareness and understanding of problems related to hazardous waste management and to protect the environment. The group studies the control of environmental stresses resulting from the generation, transportation, storage, treatment, and disposal of hazardous waste.

Activities: HWF encourages cooperation of private enterprise and industrial, governmental, and research organizations regarding hazardous waste management. The organization discusses issues surrounding hazardous waste management, transportation, oil, storage tanks, and water and groundwater pollution, as well as management of environmental pollutants and related issues. The group maintains a speakers' bureau and holds annual meetings.

Publications: None

Hazardous Waste Treatment Council (HWTC)
1440 New York Avenue, NW, Suite 310
Washington, DC 20005

Description: Founded in 1982, the council has 60 members and a staff of 7. It includes firms interested in the use of high-technology treatment in the management of hazardous waste and restricting use of land disposal facilities to protect humans and the environment.

Purpose: HWTC seeks minimization of hazardous waste and use of alternative technologies in its treatment, such as neutralization, fixation, reclamation, recycling, and incineration.

Activities: The council encourages land disposal prohibitions, advocates reduction in the amount of hazardous waste produced annually and expansion of the EPA hazardous waste list, and promotes use of treatment technologies as a cost-effective approach to Superfund site cleanups. It works with state, national, and international officials to develop programs utilizing treatment and minimizing land disposal; holds technical workshops and seminars; and participates in federal legislation, litigation, and regulatory development. The council maintains a library and a speakers' bureau, compiles statistics and a mailing list, and provides telecommunication services. The group holds annual meetings with exhibits.

Publications: Conference Proceedings (annual); *Membership Capability Profiles*

National Association of Solvent Recyclers (NASR)
1333 New Hampshire Avenue, NW, Suite 1100
Washington, DC 20036

Description: Founded in 1980, this association has 70 members and a staff of 3. Members are firms interested in recycling and reclaiming industrial solvents.

Purpose: The association promotes recycling and making useful items that might otherwise be wasted.

Activities: The association uses discarded materials for industrial fuels to save on energy, monitors and reports on legislative action regarding solvent recycling, and compiles statistics on the subject. Three committees function within the association: Insurance, Governmental Regulations, and Technology and Safety. The group holds semiannual conferences.

Publications: Flashpoint (periodic); *NASR Membership List* (semiannual)

National Resource Recovery Association (NRRA)
1620 I Street, NW
Washington, DC 20006

Description: Founded in 1982, this association has 250 members and staff of 3, with a budget of $50,000. It is affiliated with the United States Conference of Mayors. Members include city, county, and state government units and public authorities and agencies involved in resource recovery; there are 175 associate members and consultants.

Purpose: The association promotes development and operation of resource recovery facilities and of district heating and cooling systems.

Activities: The association develops recycling and urban waste energy systems to process and burn solid waste to produce steam or electricity to be used by industry, utilities, and private users. It also acts as a forum for exchange of information, offers professional and technical services, holds seminars, and monitors legislative activities. The group holds annual meetings.

Publications: City Currents (quarterly); *Directory of Associate Members* (annual); *Information Bulletin* (periodic)

Nuclear Waste Project (NWP)
c/o Environmental Policy Institute
218 D Street, SE
Washington, DC 20003

Description: This is a project of the Environmental Policy Institute.

Purpose: The project researches and analyzes federal nuclear waste policies, studying how they change and how they affect the management of commercial and military nuclear waste.

Activities: NWP coordinates citizen involvement in enforcing federal uranium mill tailings disposal legislation, and helps citizens take part in government hearings and rule making. The project advocates strong regulation of uranium mills and monitors the Mill Tailings Radiation Control Act of 1978.

Publications: None

Secondary Lead Smelters Association (SLSA)
6000 Lake Forest Drive, Suite 350
Atlanta, GA 30328

Description: Founded in 1976, this association has 45 members and 1 staff. Members are recyclers of lead, oxide manufacturers, consulting companies, and industry equipment suppliers.

Purpose: SLSA furnishes information regarding safety and environmental control, makes industry studies, offers safety and health program recommendations to eliminate hazards to employees, and conducts research programs. The organization also compiles statistics on industrywide programs. The group meets three times a year.

Publications: Newsletter (bimonthly); meeting minutes

International Organizations

Dyes Environmental and Toxicology Organization
Clarastrasse 4
CH-4005
Basel 5, Switzerland

Description: Founded in 1974, this organization has 28 members and a staff of 7, with a budget of $1 million. It consists of two regional groups, and uses English language. Membership is multinational; it is composed of firms from eight countries who manufacture synthetic organic dyestuffs and pigments.

Purpose: The organization provides protection for those who use dyestuffs and pigments and gives assistance to government departments and agencies concerned with the ecotoxicological impact of these products.

Activities: The organization coordinates efforts of members to minimize environmental damage from the use and application of dyestuffs and pigments, and conducts analytical studies on the toxicology and ecology of these products and makes recommendations. Committees include Pigments Advisory Committee, Technical Committee, and U.S. Operations Committee; and subcommittees on analytical ecology, labeling, and regulations and toxicology. The organization maintains a 400-volume library and holds annual meetings with seminars.

Publications: Annual Report; safe-handling guidelines and research articles

European Association of Poison Control Centers (EAPCC)
1, Rue Joseph Stallaert, 15
B-1060 Brussels, Belgium

Description: Founded in 1964, this association has 269 members. It conducts business in French and corresponds in English. The multinational membership includes scientists working in clinical toxicology and related fields.

Purpose: EAPCC aims to improve contacts among toxicologists.

Activities: The association represents clinical toxicologists at international organizations such as the World Health Organization and the Commission of European Communities. It provides telecommunication services and holds biennial congresses.

Publications: Newsletter (in English, quarterly)

European Chemical Industry Ecology and Toxicology Centre (ECETOC)
250 Avenue Louise, Boite 63
B-1050 Brussels, Belgium

Description: Founded in 1977, this group has 58 members and staff of 9, with a budget of $1.4 million. The multinational membership includes companies and individuals in 13 countries that are active in the chemical industry. Business is conducted in English.

Purpose: ECETOC strives to help reduce the ecological impact of the manufacture, processing, and use of chemicals.

Activities: The center offers ecological and toxicological information to those who desire it, develops methods for handling of chemicals from toxicological and ecological viewpoints, and cooperates in finding solutions for those who have problems related to chemical use. The center also conducts literature studies and experimental testing programs, maintains a task force, and provides telecommunication services. The group holds periodic meetings.

Publications: Information Sheet (monthly); monographs, technical reports, and documents

The International Association of Forensic Toxicologists (Toxicology) (TIAFT)
c/o Neville Dunnett
Soham House
Snailwell Road
P.O. Box 15
Newmarket, Suffolk CB7DT
England

Description: Founded in 1963, this association has a multinational membership of 1,000. Business is conducted in English.

Purpose: TIAFT cooperates with and coordinates efforts among forensic toxicologists.

Activities: The association encourages research in the field of chemical and systematic toxicology. It holds annual conferences.

Publications: Bulletin of TIAFT (3/year)

International Association of Medicine and Biology of Environment (IAMBE)
115, Rue de la Pompe
F-75116 Paris, France

Description: This association has a budget of $2 million. It works in French and Spanish, and corresponds in English. Membership is multinational, and includes individuals, associations, and firms in 72 countries interested in ecological medicine and biology.

Purpose: IAMBE promotes the study of how humans adapt to the environment and the study and treatment of sicknesses arising from this adaptation.

Activities: The association collects and offers information concerning the protection of humankind and the environment. It makes contact easy among people who deal professionally with problems related to the protection of humans and their surroundings by organizing symposia and congresses, conducting seminars and courses, and providing telecommunication services. The association has many committees, one of which is concerned with automotive and industrial effluence and its effects on health. The group holds periodic meetings known as World Conferences on Hazardous Waste (with exhibits).

Publications: None

World Federation of Associations of Clinical Toxicology Centers and Poison Control Centers
c/o CIRC
150, Cours Albert-Thomas
F-69362 Lyon, France

Description: Founded in 1975, this group has 37 members. It is multinational and uses French, Russian, and Spanish, and corresponds in English. It consists of associations of poison control centers, national poison control centers, and national and international organizations dealing with toxicology.

Purpose: The organization assists developing countries in toxicology education and training, collects and disseminates information, and encourages prevention through control of products and their distribution and consumer information and education.

Activities: The organization acts as a liaison with the World Health Organization. It compiles statistics, maintains a 200-volume library, and organizes working groups. The group holds quadrennial meetings.

Publications: Bulletin of the World Federation (quarterly); *Collection de Medecine Legale et Toxicologie Medicale* (report of meeting proceedings, periodic); *Journal de Toxicologie Medicale* (monthly); *Catastrophies Toxiques* and other books; membership list (periodic)

World Meteorological Organization (WMO)
Case postale No. 5 CH-1211
Geneva 20, Switzerland

Description: This is a specialized agency of the United Nations, with a membership of 1,716 countries and territories. Its predecessor, the International Meteorological Organization (IMO), was organized in 1878. The organization helps to facilitate worldwide cooperation in the establishment of networks of stations for making meteorological, hydrological, and other geophysical observations and promotes the establishment of meteorological centers.

Activities: Two recently established programs of the WMO consider pollution of the atmosphere and climactic change. Since the Chernobyl nuclear power plant accident in 1986, WMO has collaborated with the International Agency for Atomic Energy to provide information relating to the transport of hazardous materials in the atmosphere. The WMO, jointly with the U.N. Environment Program, established the Intergovernmental Panel on Climate Change in 1988 to assess available scientific information on climate warming, to assess the environmental and socioeconomic impacts of climate warming, and to formulate international response strategies.

Publications: None

Sources for Chapter 4

Encyclopedia of Associations, Vol. 1. *National Organizations of the U.S.* Detroit, MI: Gale Research Co., 1990.

Encyclopedia of Associations, Vol. 4. *International Associations.* Detroit, MI: Gale Research Co., 1990.

Encyclopedia of Governmental Advisory Organizations, 1988–1989, 6th ed. Detroit, MI: Gale Research Co., 1987.

The United States Government Manual, 1989/90. Washington, DC: U.S. Government Printing Office, 1989.

5

Bibliography

ALTHOUGH STUDIES OF ENVIRONMENTAL PESTICIDES were undertaken more than 2,000 years ago, the modern recognition of the pollution of the environment and its consequences to humanity began in the 1960s, with the publication of *Silent Spring,* by Rachel Carson. Environmental pollution by hazardous material and toxic waste is now recognized by the public as one of the nation's major problems. Intrinsic to this concern is not only the cleanup of the devastation of the past but the need to preserve our environment in the future. As a response to these endeavors there has been a massive increase in the literature. Although most of the publications are technical, intended to provide solutions to the environmental problem, many studies consider human aspects of the issue. This selected bibliography provides a wide perspective on the problems connected with environmental pollution and their possible solutions.

Reference Sources

Dictionaries and Indexes

Berger, Melvin. **Hazardous Substances: A Reference.** Hillside, NJ: Enslow, 1986. 128p. ISBN 0-89490-116-8.

People who use such substances as asbestos, turpentine, kepone, and lindane as insecticides or mica and talc in factories will find this guide most useful. Every laboratory should have a copy to show the hazards of benzene, formaldehyde, and sulfuric acid. Terms are defined in such a way that the book is useful to anyone. The author gives acronyms, chemical formulas, and phsical descriptions of substances. Both Bhopal, India, and Love Canal chemicals are mentioned. The reference books and pamphlets cited by the author are an excellent source for additional reading.

Hodgson, Ernest, Richard B. Mailman, and Janice E. Chambers. **Dictionary of Toxicology.** New York: Van Nostrand Reinhold, 1988. 395p. ISBN 0-442-31842-1.

This is a most useful dictionary of terms used in the field of toxicology. About 60 contributors worked on this book, which includes terminology for toxic chemicals, measurement of toxicity, environmental pollutants, regulations, and organizations and government bodies in North America and Western Europe. Anatomical, pathological, biochemical, and physiological terms related to toxicology are included. Structures of compounds and diagrams are given. The dictionary should be of value to undergraduate and graduate students and scientists in areas other than toxicology.

Norback, Craig T., and Judith C. Norback, eds. **Hazardous Chemicals on File.** New York: Facts on File, 1988. 3 vols. ISBN 0-8160-1353-5.

This book was written to be used by the general public for information on how to protect themselves against hazardous chemicals. There are 327 main entries listing toxic chemicals or groups of related substances. Entries listing chemical names, systematic names, and trade names are in alphabetical order. Types of hazardous information such as permissible exposure limit, monitoring and measurement procedures, health hazards from exposure, methods for personal protection, waste disposal methods, and leak and spill procedures are included in the book. The main entries are preceded by an introduction and background on OSHA, NIOSH, and EPA. Addresses and telephone numbers of OSHA are provided. A pamphlet gives an alphabetical substance list and synonyms. The format allows for quick access to information as needed.

Suspect Chemicals Sourcebook, 1985. 4th ed. Prepared with the cooperation of the Synthetic Organic Chemical Manufacturers Association by Roytech Publications, 1985. Burlingame, CA: 1985. 1 vol. (various paging). ISBN 0-9612092-1-6.

The fourth edition brings together 6,000 chemical names from 21 sources, which include the National Toxicology Program, NIOSH special hazard reviews, and the Toxic Substances Control Act. Complete directions are given on how to use the book with its long indexes. The major indexes are Part I, Chemical Name Index; Part II, CAS Registry Number Index; Part III, OSHA Chemical Hazard Communication Standard—Chemical Names Index; and Part IV, OSHA Chemical Hazard Communication Standard—CAS Registry Number Index. Each index gives the CAS number, the chemical name, and a reference number to the source list. Indexes are followed by appendices.

Appendix I gives a summary of the OSHA Chemical Hazard Communication Standard Covering Chemicals in the Workplace, Hazard Evaluation, Labels and Other Forms of Warning, Trade Secrets, etc. Appendix II gives the text of OSHA Chemical Hazard Communication Standard. Appendix III presents the *Federal Register,* November 25, 1983, concerning the history, overview, and regulatory analysis of the OSHA Chemical Hazard Communication Standard. This book is a valuable addition to libraries; it will be useful to chemical companies and especially chemists who work with hazardous chemicals and are concerned about their safety.

Bibliographies

Brown, Ronald L. **"Marine Oil Pollution Literature: An Annotated Bibliography."** *Journal of Maritime Law and Commerce* 13 (April 1982): 373–390.

This is an annotated bibliography listing material on the environmental, economic, and social impact of marine oil pollution.

Champ, Michael A., and P. Kilho Park. **Global Marine Pollution Bibliography: Ocean Dumping of Municipal and Industrial Wastes.** New York: Plenum Press, 1982. 339p. ISBN 0-306-65205-6.

This bibliography was compiled as a result of growing global concern with the growing problem of waste disposal. It focuses on municipal and industrial wastes dumped at sea. Some of the 1,742 items are annotated. The bibliography has a subject and author approach to the citations. There is also a list of abbreviations of agencies/institutions with addresses.

Harnly, Caroline D. **Agent Orange and Vietnam: An Annotated Bibliography.** Metuchen, NJ: Scarecrow Press, 1988. 401p. ISBN 0-8108-2174-5.

This annotated bibliography is divided into six sections: general articles, ethical and political issues, effects on Vietnam's ecology, health effects and costs of exposure on the Vietnamese, disposal of leftover Agent Orange, and finally effects on the Vietnam veteran. The book has an author index.

Hazardous Waste Bibliography. OSWER Directive, 9380.1-02. Washington, DC: U.S. Environmental Protection Agency, Office of Solid Waste and Emergency Response, 1987. 42p. No ISBN.

This bibliography includes materials relating to the cleanup of hazardous wastes that should be available to federal and state hazardous waste staffs and their libraries. Some of the documents are technical, but most useful for cleanup. It is divided into four parts—"A Prime" list, "A" list, "B" list, and "C" list, ranging from those documents most critically important to those considered to be useful.

Louden, Louise. **Toxicity of Chemicals and Pulping Wastes to Fish.** Bibliographic Series, no. 265, suppl. I. Appleton, WI: Institute of Paper Chemistry, 1979. 425p. ISBN 0-686-63984-7.

More than 1,600 government reports, dissertations, articles, books, and foreign works are listed, with complete bibliographic information and arranged in alphabetical order by authors. This is a good source for available literature.

Magorian, Christopher. **Public Participation and Hazardous Waste Facility Siting: An Annotated Bibliography.** Philadelphia, PA: Pennsylvania Environmental Research Foundation, 1982. 75p. No ISBN.

The purposes of this bibliography are to inform people about issues involved in the siting of hazardous waste facilities, to encourage their participation in the siting process, and to develop solutions to hazardous management problems. The first part of the bibliography lists literature giving an overview of hazardous waste issues; the second part deals with public participation in environmental decisions; the third section deals with facility siting; and lastly technical information is given. Literature cited covers 1979–1981.

Morales, Leslie Anderson. **Toxic Dumping in the Third World: A Bibliography.** Public Administration Series, bibl. no. P2781. Monticello, IL: Vance Bibliographies, 1989. 7p. ISBN 0-7920-0391-8.

Since Western Europe exported waste to Africa and the United States exported waste to the Caribbean and Latin American countries, these Third World nations are questioning the ethics and economics of the dumping of such hazardous material. This short bibliography contains articles and monographs published by the United Nations and the U.S. Congress as well as journal articles.

Vance, Mary. **Chemical Plant Wastes: A Bibliography.** Public Administration Series, bibl. no. P2749. Monticello, IL: Vance Bibliographies, 1989. 20p. ISBN 0-7920-0329-2.

Vance has compiled an alphabetical (by author and title) list of journal articles dealing with wastes from chemical plants and their disposal.

————. **Industrial Waste Disposal: A Bibliography.** Public Administration Series, bibl. no. P959. Monticello, IL: Vance Bibliographies, 1982. 53p. No ISBN.

This bibliography begins with a list of books and government documents on industrial waste disposal. It is followed by an alphabetical list of journals in the subject area and articles found in each journal dealing with the disposal of industrial waste.

Books

General

Armstrong, Neal E., and Akira Kudo, eds. **Toxic Materials: Methods for Control.** Water Resources Symposium, no. 10. Austin: Center for Research in Water Resources, University of Texas. 1983. 380p. No ISBN.

Toxic waste has become a major environmental problem because of past disposal practices, which did not contain toxic material well enough to prevent it from invading our environment. Toxic waste has contaminated the soil in which our crops grow, the air we breathe, our drinking water, and shellfish, and requires more stringent controls. Two major pieces of legislation that regulate toxic material in liquid effluents and solid wastes are the Water Pollution Control Act of 1972 and the Resource Conservation and Recovery Act of 1976. These laws limit the discharge of toxic materials and provide ways to handle hazardous wastes. This book is a compilation of papers presented by experts at a symposium for controlling toxic materials that was held on the campus of the University of Texas at Austin, May 4–6, 1982. Such topics as federal, state, industrial, and ethical views of toxic control, disposal, and control of radioactivity were discussed.

Bennett, Gary F., Frank S. Feates, and Ira Wilder, eds. **Hazardous Materials Spills Handbook.** New York: McGraw-Hill, 1982. 1 vol. (various paging). ISBN 0-07-004680-8.

This handbook is an excellent source to use to examine procedures for dealing with dangerous spills, equipment needed, information on specific chemicals, and methods of responding to spills. Both the political and legal aspect of spills are discussed. The book covers problems of working with volatile chemicals, such as monitoring, controlling, safe evacuation, and evaporation, and so should be available to anyone working with hazardous spills.

Brunner, David L., Will Miller, and Nan Stockholm, eds. **Corporations and the Environment: How Should Decisions Be Made?** Los Altos, CA: Committee on Corporate Responsibility, 1980. 176p. ISBN 0-96576-020-9.

This book describes the actions that took place at the Symposium on Corporate Environmental Decision Making, conducted in 1980 at the Stanford Graduate School of Business. Participants discussed such problems as the urgency of the environmental problem, how to deal with it, how environmental constraints affect businesses, and how sound environmental decisions can be formulated and implemented. Specific case studies illustrating difficult environmental situations confronted by three major corporations—Bethlehem Steel Corporation, 3M Company, and AMAX Incorporated—are examined. Successful corporations must overcome the problems of environmental protection and regulatory reform confronting everyone. Government and industry must cooperate. The book includes a list of suggested readings.

Cooper, M. G., ed. **Risk: Man-Made Hazards to Man.** New York: Oxford University Press, 1985. 141p. ISBN 0-19-854155-4.

This book is based on eight lectures presented at Wolfson College, Oxford, in the spring of 1984. Assessment and management of environmental health risks posed by industry and other sources are discussed. Exposure to human-made radiation from nuclear power always comes to the fore in discussions of risk because of its links with the atomic bomb, cancer, birth defects, and so on.

Davis, Charles E., and James P. Lester, eds. **Dimensions of Hazardous Waste: Politics and Policy.** Contributions in Political Science, no. 200. Westport, CT: Greenwood Press, 1988. 256p. ISBN 0-313-25988-5.

This volume, prepared under the auspices of the Policy Studies Organization, has many contributors who express their views on siting and disposal policy processes at local, state, and federal levels and also in Western Europe and along the U.S.-Mexico border. The book has figures, tables, and a selected bibliography. Short biographical sketches of the contributors are also of interest.

Epstein, Samuel S., Lester O. Brown, and Carl Pope. **Hazardous Waste in America.** San Francisco: Sierra Club Books, 1982. 593p. ISBN 0-87156-294-4.

Hazardous waste problems have increased over the years. More than 80 billion pounds of toxic waste are disposed of in the United States each year, and the volume keeps growing. Also, many sites are being discovered that have been used indiscriminately by waste dumpers that add to the problems. The chemical industry has added new chemical wastes in large quantities, which increases the disposal problem. This book presents a blend of the economics and politics as well as the history of toxic waste and the damage done to the environment and public health. There are 14 appendices giving such information as estimated industrial hazardous waste generated by region, toxicological effects of hazardous chemicals found at Love Canal, sites subject to legal action by the Hazardous Waste Enforcement Effort of the EPA, federal statutes governing toxic wastes, the EPA's list of potential hazardous waste disposal sites, and ranking of states by volume of hazardous waste generated.

Freeman, Harry M., ed. **Standard Handbook of Hazardous Waste Treatment and Disposal.** New York: McGraw-Hill, 1989. 1 vol. (various paging). ISBN 0-07-022042-5.

This handbook gives a nontechnical introduction to the Resource Conservation and Recovery Act (1976) and the disposal and elimination of hazardous material. It lists vendors of waste disposal systems, with their addresses and telephone numbers and the technologies they use. There is an alphabetical list of symbols and abbreviations used in the book. The index is quite detailed.

Freudenberg, Nicholas. **Not in Our Backyards!** New York: Monthly Review Press, 1984. 304p. ISBN 0-85345-653-4.

People of all economic levels have fear and concern about how our country manages hazardous waste. When their communities are chosen for hazardous waste sites, they are disturbed. No one wants a Love Canal or Bhopal in their backyard. Communities have become concerned and willing to fight because public officials and policymakers seem to be trying to keep people calm by playing down the seriousness of the problem rather than doing something about it. Waste handlers are interested in the profits they make. The book begins by describing how government agencies and corporations pollute the air, water, food, and soil with toxic substances and then discusses actions communities can take against those threats to their health and environment. It goes on to tell how to organize groups and press legal action. Each chapter has at least one case history. References are given for each chapter.

Harrison, L. Lee, ed. **The McGraw-Hill Environmental Auditing Handbook: A Guide to Corporate and Environmental Risk Management.** New York: McGraw-Hill, 1984. 1 vol. ISBN 0-07-026859-2.

This book is divided into five parts. Part 1 explains what environmental auditing is and discusses ways in which it will be helpful to industry. Part 2 discusses environmental risks, the Toxic Substances Control Act, Superfund, and the like. Part 3 tells how to undertake an environmental audit and gives several case studies—Atlantic Richfield, General Motors Corporation, Olin Corporation, and Pennsylvania Power and Light Company. Part 4 explains the confidentiality of the audit. Part 5 gives the EPA's view of environmental auditing.

Huisingh, Donald, and Vicki Bailey, eds. **Making Pollution Prevention Pay: Ecology with Economy as Policy.** New York: Pergamon Press, 1982. 156p. ISBN 0-08-029417-0.

This book contains papers given at the Phase 1 Symposium on Philosophy, Technology, and Economics of Pollution Prevention held May 26–27, 1982, at the W. C. Benton, Jr., Convention Center in Winston-Salem, North Carolina. The symposium was sponsored by the Mary Reynolds Babcock Foundation and was attended by industrial, governmental, academic, and private citizens interested in pollution prevention. "Pollution Prevention Pays" is a part of the state of North Carolina's plan for controlling toxic substances and hazardous waste management. Environmental quality will be improved by reducing or not releasing hazardous substances into the environment. By making pollution prevention pay, industry, government, and citizens are motivated to work together to make prevention work.

Hynes, H. Patricia. **The Recurring Silent Spring.** The Athene Series. New York: Pergamon Press, 1989. 225p. ISBN 0-08-037117-5.

This book is organized around the basic questions that are asked of the inspired work that fueled the modern environmental movement. In the first chapter the main aim is to tell the story of how Rachel Carson came to write *Silent Spring*. Chapter 2 presents the major ideas of *Silent Spring* and analyzes the work for its prophetic, consciousness-raising, political, and visionary qualities. Chapter 3, "The World Aroused," chronicles and analyzes the vortex of industrial and government reaction to *Silent Spring*. Chapter 4 is a critical examination of the Environmental Protection Agency, now generally recognized as the most prominent environmental agency in the world. The final chapter considers the health problems of women created by chemicals. The basic purpose of this chapter is to hold the EPA to the standard passion and politics set by Rachel Carson. This volume's major purpose is to demonstrate the wisdom revealed by Rachel Carson in *Silent Spring*.

Ives, Jane H., ed. **The Export of Hazard: Transnational Corporations and Environmental Control Issues.** Boston: Routledge & Kegan Paul Ltd., 1985. 229p. ISBN 0-7102-0072-2.

Ralph Nader wrote the foreword to this book. Pesticides that cannot be sold in the United States can be and are sold to other countries and returned to us on imported food. Exporting hazardous material can often cause foreign incidents. Questions involving abuses in international commerce are raised in the book. There are double standards in the control of industrial hazards; one country may ban a dangerous practice or product, while others do not. Rashid Shaikh prepared a bibliography for the book on the export of hazardous products from the United States. In the appendix the Bhopal disaster is used as a case study in double standards.

Kiang, Y., and A. A. Metry. **Hazardous Waste Processing Technology.** Ann Arbor, MI: Ann Arbor Science Publishers, Inc., 1982. 549p. ISBN 0-250-40411-7.

The authors intended this to be used as a reference book by professionals involved with hazardous waste activities. The book is divided into two parts. Part 1 gives an overall view of hazardous waste management and regulations, classification of hazardous waste, incineration handling of waste, and thermal processes. Part 2 discusses disposal site selection and chemical and biological treatment processes for disposal of hazardous waste. References are found at the end of each chapter. Diagrams and tables are included in the book.

Kiefer, Irene. **Poisoned Land: The Problem of Hazardous Waste.** New York: Atheneum, 1983. 90p. ISBN 0-689-30837-X.

Hazardous waste is hazardous because it can ignite, corrode, react, or explode and infect people and the environment. Much of this book is devoted to the

Love Canal episode, which was one of the worst environmental disasters ever to happen. Such mistakes can be prevented in the future by minimizing the amount of hazardous waste produced and by enforcing RCRA regulations more strictly. The book has a glossary.

Lave, Lester B., and Arthur C. Upton, eds. **Toxic Chemicals, Health, and the Environment.** Baltimore, MD: Johns Hopkins University Press, 1987. 304p. ISBN 0-8018-3473-2.

This book is the product of a forum, held at the New York City Rene Dubos Center for Human Environment in 1984, on toxic chemicals. A main theme from the forum was that there should be better information developed and disseminated among our leaders regarding toxic chemicals' effects on human health and the environment. We must learn to control and minimize the production of toxic chemicals and control our exposure to them. The book is concerned with the management of environmental toxic chemicals without affecting our industrial economy. An excellent chapter deals with our exposure to toxic chemicals through the food we eat, and another lengthy chapter discusses the cleanup of contaminated sites. A list of contributors is found at the beginning of the book.

Leonard, H. Jeffrey. **Pollution and the Struggle for the World Product: Multinational Corporations, Environment and International Comparative Advantage.** New York: Cambridge University Press, 1988. 254p. ISBN 0-521-34042-X.

This book investigates the problem of whether or not standards for pollution control have pushed U.S. industries to relocate abroad. It also looks into the effects regulations have on underindustrialized countries. This study deals with public concern about industrial pollution and pollution control regulations. The book includes many tables.

Majumdar, Shyamal K., and E. Willard Miller, eds. **Hazardous and Toxic Wastes: Technology, Management and Health Effects.** Easton, PA: Pennsylvania Academy of Science, 1984. 442p. ISBN 0-9606670-2-4.

The problem of the disposal of the various types of wastes created by an industrial society is not a new one, but one that has reached a critical dimension in our time. Modern industrial wastes not only can devastate the natural environment, but have the potential to affect the health of millions of human beings. This volume is divided into five parts. Part 1 considers waste types and treatment and disposal methods. The sites of hazardous and toxic wastes, as to their distribution, selection, and geological considerations, are discussed in Part 2. Part 3 covers transportation, emergency response, and preparations needed in a toxic spill emergency. Part 4 includes chapters on management, regulations, and economic considerations, and the final part is devoted to environmental and health effects of hazardous wastes.

Petulla, Joseph M. **Environmental Protection in the United States: Industry, Agencies, Environmentalists.** San Francisco: San Francisco Study Center, 1987. ISBN 0-936434-21-X.

The book begins with the history of environmental protection. We are still threatened by the technologies that generate nuclear and toxic wastes that contaminate both surface and groundwater and the air. Environmental regulations are made by the government but are not always followed closely by industrial leaders if they can delay enactment. EPA priorities are often determined by the most recent environmental disaster, such as the Bhopal accident and oil spills. The laws we already have must be enforced more effectively. A bibliography is found at the end of each chapter. Several excellent tables illustrate the book, such as "Attitudes on Environmental Concerns Early 1980s," "Trends in Industrial Commitment," "How Much Are Future Benefits Worth," and "Elements of Risk Assessment and Risk Management."

Piasecki, Bruce, ed. **Beyond Dumping: New Strategies for Controlling Toxic Contamination.** Westport, CT: Quorum Books, 1984. 239p. ISBN 0-89930-056-1.

Many authors contributed chapters to make this book possible. In the first section, the authors discuss factors sustaining the crisis of dumping toxic waste, such as limitations of landfills and difficulties faced in efforts by the EPA to stop the growing problem. The EPA is often delayed in its actions by political decisions and toxic waste regulations. Parts 2 and 3 tackle ways to control the dumping of toxic waste, with emphasis on the part played by industry. Detoxification as an alternative to dumping is discussed. Appendix A gives some alternative technologies for hazardous waste management. Appendix B discusses landfills of the future, and Appendix C discusses wastes and new solutions and impediments encountered. The book has an annotated bibliography and lists related magazine articles.

Postel, Sandra. **Altering the Earth's Chemistry: Assessing the Risks.** Worldwatch Paper 71. Washington, DC: Worldwatch Institute, 1986. 66p. ISBN 0-916468-72-0.

Human activities have altered the earth's chemistry, and these changes may result in ecological and economic consequences, such as risks to food security, to forests, and to human health. In industrial countries people are exposed to harmful pollution. Pollution from metals and fossil fuels are toxic to humans. Chemicals are introduced to the environment with no knowledge as to the effects they will have on the atmosphere, water, soil, food supplies, forests, and human health.

————. **Defusing the Toxics Threat: Controlling Pesticides and Industrial Waste.** Worldwatch Paper 79. Washington, DC: Worldwatch Institute, 1987. 69p. ISBN 0-916468-80-1.

After discovering the insecticidal properties of DDT, Paul Müller, a Swiss, received the Nobel Prize in medicine in 1948. But DDT is now banned or restricted in its use in many countries. Chemical pesticides are used more extensively in industrial countries than in developing countries. It seems the method of managing industrial waste is more or less "out of sight, out of mind." Laws to curb land disposal of hazardous waste were passed in 1984, but the EPA delayed their implementation because of lack of facilities to treat and handle waste in a safer way. Hazardous waste is discharged into sewers, rivers, and streams. Some is recycled, detoxified, or destroyed before it reaches the environment. The author discusses attempts made by China, Brazil, the United States, Nicaragua, and other countries to break the pesticide habit. She concludes the book with a discussion of how to reduce pesticide use in agriculture as well as the amount of industrial waste.

Regenstein, Lewis. **America the Poisoned: How Deadly Chemicals Are Destroying Our Environment, Our Wildlife, Ourselves and How We Can Survive!** Washington, DC: Acropolis Books, 1982. 414p. ISBN 0-87491-486-8.

Regenstein tells the story of how chemicals are killing millions of Americans. The EPA has called this toxic chemical contamination one of the most serious problems we have ever faced. Toxic substances that cause cancer, birth defects, and nerve and brain damage are a potential threat to the life and health of everyone and impossible to avoid. Toxic substances, including carcinogens, are in the air we breathe, the food we eat, the water we drink, and the consumer products we use. Our environment is contaminated with synthetic pollutants that endanger our lives. Many banned chemicals are shipped abroad, legally, where they poison people in foreign countries and return to us as residues on tea, coffee, sugar, bananas, and so on, which we import. Even so, the chemical industry spends millions of dollars on advertising that denies there is a problem. More than 20 years ago Rachel Carson wrote *Silent Spring*, warning us of the hazards of toxic chemicals, but they are still used, and many new chemicals are introduced each year.

Sax, N. Irving, and Richard J. Lewis, Sr. **Dangerous Properties of Industrial Materials.** 7th ed. New York: Van Nostrand Reinhold, 1989. 3 vols. ISBN 0-442-28020-3.

This book was first published in 1951 as *Handbook of Dangerous Materials* and was the reference source to the hazardous chemical environment. This edition has been rewritten to include data from chemical and medical literature, and also state and federal standards and ratings. More than 18,000 chemicals and 40,000 synonyms are included. This edition is divided into eight parts, covering toxicology; industrial air contaminant control; industrial and environmental cancer risks; occupational biohazards; nuclear medicine applications, benefits, and risks; an alphabetical list of chemicals and data on identity and toxicity; a list of synonyms; and a CODEN index. The index gives CAS, RN, and NIOSH numbers, so that the data can be found in any online database.

Sherry, Susan, and others. **High Tech and Toxics: A Guide for Local Communities.** Sacramento, CA: Golden Empire Health Planning Center, 1985. 467p. ISBN 0-89788-090-0.

The purpose of this book is to show the role a community can play in designing preventive programs to protect the public from industrial chemical exposure. Industry benefits as well as the public in the prevention of exposure to toxic chemicals, because of the reduction of potential liability and the need for further regulation, as well as the favorable economics associated with the recycling of and need for industrial chemicals. Government policies to minimize chemical risks need to be developed while high-tech industries are still young. Communities should study the mistakes that led to toxic incidents in the past so that they do not occur again. Industry must assume responsibility for the hazardous material it uses and for storage and disposal problems.

Sittig, Marshall. **Handbook of Toxic and Hazardous Chemicals and Carcinogens.** 2d ed. Park Ridge, NJ: Noyes Publications, 1985. 950p. ISBN 0-8155-1009-8.

Toxicological and chemical information about hazardous waste and industrial chemicals is given. About 800 compounds are covered, drawn from the EPA, the American Conference of Government Industrial Hygienists (ACGIH), the National Institute of Occupational Safety and Health (NIOSH), the U.S. National Toxicology Program, the Dutch Chemical Industry Association, and other sources. Description of each chemical compound includes color, odor, and melting or boiling point. Entries are in alphabetical order, making the book easy to use. A very lengthy bibliography is a helpful resource.

Tryens, Jeffrey, ed. **The Toxics Crisis: What the States Should Do.** Washington, DC: Conference on Alternative State and Local Policies, 1983. 105p. ISBN 0-89788-072-2.

This book was written by several authors who are knowledgeable about the issues related to toxic chemicals. The book investigates problems and solutions in four major areas—protecting workers, protecting the community, waste management, and waste cleanup. Most states must find new ways to finance toxic waste programs, to develop and use scientific expertise, and to try to get the public more involved.

Webster, James K., ed. **Toxic and Hazardous Materials: A Sourcebook and Guide to Information Sources.** Bibliographies and Indexes in Science and Technology, no. 2. Westport, CT: Greenwood Press, 1987. 431p. ISBN 0-313-24575-4.

This bibliography covers various types of publications, indexes, and databases on hazardous materials. It also contains a list of associations, government agencies, and research organizations in the field.

Whelan, Elizabeth M. **Toxic Terror.** Ottawa, IL: Jameson Books, 1985. 348p. ISBN 0-915463-09-1.

The author discusses the impact of a variety of chemicals and technologies, pesticides, PCBs, PBBs, dioxin, and formaldehyde on human health and the environment. In the appendix she discusses the myth of a cancer epidemic.

Management

Bhatt, H. G., Robert M. Sykes, and Thomas L. Sweeney, eds. **Management of Toxic and Hazardous Wastes.** Chelsea, MI: Lewis Publishers, 1986. 418p. ISBN 0-87371-023-1.

This book was compiled as the result of the Third Ohio Environmental Engineering Conference, held in Columbus, Ohio, in 1983, but the material was updated before publication. Some of the topics treated are solid waste facility siting, industry's view of hazardous waste management, impact on groundwater contamination control and treatment, land disposal, disposal site cleanup, and legal considerations.

Cashman, John R. **Management of Hazardous Waste: Treatment/Storage/ Disposal Facilities.** Lancaster, PA: Technomic Publishing Company, 1986. 311p. ISBN 0-87762-453-4.

To get material for this book the author visited many firms to study how they managed treatment, disposal, and storage facilities for hazardous waste. He relates here what hazardous waste facility managers are attempting to do and how they go about it. The author makes no attempt to compare the procedures used by one firm with the procedures of another firm. However, he does use specific firms as examples in each chapter. The last chapter lists sources of information, resource materials, agencies, organizations and associations, and periodical publications related to hazardous waste management.

Chatterji, Manas, ed. **Hazardous Materials Disposal: Siting and Management.** Brookfield, VT: Gower Publishing Company, 1987. 331p. ISBN 0-566-05293-8.

This volume is divided into two parts. The first deals with issues in the location and management of unwelcome facilities. This issue of unwelcome facilities has become significant since the Bhopal, Chernobyl, and Three Mile Island accidents. Environmental pollution is caused by hazardous waste sites that are found in the United States. The sites must be studied if we are to assess the risk and prevent the spread of contamination. The growth of synthetic chemical industries has caused environmental risk assessment to increase. The second part of the book deals with location-allocation modeling of hazardous facilities. Tables and graphs are included, along with a selected bibliography on location-allocation modeling and references at the end of chapters.

Council on Economic Priorities. **Hazardous Waste Management: Reducing the Risks.** Covelo, CA: Island Press, 1986. 316p. ISBN 0-933280-31-9.

The Council on Economic Priorities studied and rated operating toxic waste disposal sites in order to help toxic waste generators determine where they might dispose of toxic wastes safely and efficiently. Chapter 4 discusses commercial hazardous waste management as a new industry. Ten toxic waste disposal sites were studied and rated, using the U.S. Environmental Protection Agency's Hazard Ranking System. The book has many excellent figures and tables and a glossary.

Dominguez, George S., and Kenneth G. Bartlett, eds. **Hazardous Waste Management, Volume 1: The Law of Toxics and Toxic Substances.** Boca Raton, FL: CRC Press, 1986. 207p. ISBN 0-8493-6356-X.

This volume was written to introduce managers to the working aspects of hazardous waste management and the laws and regulations governing it. The book discusses hazardous waste generation, storage, treatment, disposal, and transportation. Legislation and remedial options of hazardous waste management are also treated. The legislative history of the RCRA is given, and an overview of the hazardous waste program under the RCRA is discussed, along with legal development of remedies for toxic tort plaintiffs and remedies for hazardous waste victims.

Hazardous Waste Management at Educational Institutions. Washington, DC: National Association of College and University Business Officers, 1987. 101p. ISBN 0-915164-36-1.

Colleges and universities became classed as generators of hazardous waste with the 1985 amendments to the Resource Conservation and Recovery Act. This book was written to give them an understanding of the hazardous waste problem and how to develop plans for the disposal of hazardous waste. The term *hazardous waste,* as used in this book, applies to chemical waste, and the book discusses its generation, treatment, storage, and disposal. Schools have disposed of hazardous waste at sites that have contaminated the environment; explosions have occurred in school laboratories that have injured students; and school buildings have been contaminated by misuse of toxic chemicals. The book is written in an easy-to-read style and includes a bibliography and a glossary.

Lindgren, Gary F. **Guide to Managing Industrial Hazardous Waste.** Woburn, MA: Butterworth Publishers, 1983. 287p. ISBN 0-250-40591-1.

The author gives an overall picture of the hazardous waste management system and defines solid and hazardous waste. He then explains fully the federal regulatory standards applied to hazardous waste generators and how some state regulations may differ from federal regulations, some even being more strict than the federal. In the third section the author discusses the development of a least-cost compliance program for companies generating hazardous waste and the use of environmental audits to evaluate the degree of compliance of the firm. The fourth section gives general information on disposal site selection, inspection, insurance options, and legal responsibility. Some examples of good management practices are described. The book has a bibliography and appendices that list EPA regional waste management offices, state solid waste agencies, examples of hazardous wastes, industrial waste exchanges, an inspection report example, and, finally, hints for hazardous waste management. Hazardous waste managers will find this a most useful book.

Majumdar, Shyamal K., E. Willard Miller, and Robert F. Schmalz, eds. **Management of Hazardous Materials and Wastes: Treatment, Minimization and Environmental Impacts.** Easton, PA: Pennsylvania Academy of Science, 1989. 473p. ISBN 0-9606670-9-1.

Humans have always thought that their environment was so vast that pollution could make only the most gradual changes. Today the problems are far more complex. The sources of hazardous and toxic wastes are many and diverse, and the damage may not only be immediate but may also have long-term effects. This volume is divided into six parts. Part 1 considers the generation, treatment, and technology of hazardous substances. Part 2 considers waste minimization and reduction methods. Part 3 covers regulations, management, and transportation. Part 4 considers the environmental and health effects of hazardous waste. Sociological factors are presented in Part 5, and Part 6 presents case studies.

Martin, Edward J., and James H. Johnson, Jr., eds. **Hazardous Waste Management Engineering.** New York: Van Nostrand Reinhold, 1987. 520p. ISBN 0-442-24439-8.

This book was written to be helpful to engineers and anyone else dealing with hazardous waste management. It gives options for the elimination or management of hazardous wastes that are most economical and also protect the health of everyone. Advantages and disadvantages of chemical, biological, and physical treatment, storage, incineration, and land disposal of wastes are discussed. One chapter is devoted to leachate management and one to siting. The book is recommended for the average citizen as well as for engineers.

Phifer, Russell W., and William R. McTigue, Jr. **Handbook of Hazardous Waste Management for Small Quantity Generators.** Chelsea, MI: Lewis Publishers, 1988. 198p. ISBN 0-87371-102-5.

The authors discuss the types of hazardous waste generated in small quantities and on-site storage and handling of such wastes. Environmental audits are necessary to be sure the generators are in compliance with regulations. Laboratories and small-quantity generators have various options for disposal of their hazardous wastes. The book has several appendices listing such things as hazardous waste agencies, solid wastes that are not hazardous, hazardous wastes, U.S. Department of Transportation definitions of hazardous waste, and federal regulations regarding hazardous waste residues in empty containers.

Pojasek, Robert B., ed. **Toxic and Hazardous Waste Disposal.** Ann Arbor, MI: Ann Arbor Science Publishers, 1979–1980. Vol. 1, 407p. ISBN 0-250-40251-3. Vol. 2, 259p. ISBN 0-250-40252-1. Vol. 3, 205p. ISBN 0-250-40253-X. Vol. 4, 313p. ISBN 0-250-40265-3.

This series of four volumes deals with various aspects of toxic and hazardous waste disposal. Volume 1 treats processes for stabilization and solidification, Volume 2 gives options for stabilization/solidification, Volume 3 shows the impact of legislation and implementation on disposal management practices, and Volume 4 discusses new and promising ultimate disposal options that must be examined if one is to choose the option best suited to the situation. The volumes contain many tables, graphs, and photos.

Proceedings of the 5th National Conference HWHM '88—Hazardous Wastes and Hazardous Materials. Silver Spring, MD: Hazardous Materials Control Research Institute, 1988. 571p. No ISBN.

This volume covers the proceedings of the 5th National Conference on Hazardous Wastes and Hazardous Materials, held April 19–21, 1988, in Las Vegas, Nevada. The conference was sponsored by the Hazardous Material Control Research Institute, and its purpose was to study the control and management of hazardous material and toxic waste, with the objective of the protection of public health, natural resources, and the environment. The RCRA Amendments of 1984 made great changes in the management of hazardous waste. A new section dealt with underground leakage from storage tanks. Permits for having a hazardous waste facility are now more difficult to obtain. These proceedings contain 130 papers and seminar outlines that stress the developments and experiences gained from RCRA and Superfund activities. A glossary contains frequently used acronyms and abbreviations.

Sarokin, David J., and others. **Cutting Chemical Wastes: What 29 Organic Chemical Plants Are Doing To Reduce Hazardous Wastes.** An INFORM Report. New York: INFORM, 1985. 535p. ISBN 0-918780-32-2.

A study was made of 29 plants by the U.S. organic chemical industry to see how their hazardous wastes could be reduced at the source of origin. The organic chemical industry is of utmost importance to everyone concerned about toxic pollution problems. The study suggested changes in processes and products and substitution of chemicals and also changes in equipment and operations of the plants to reduce the hazardous wastes. The chemical industry's products—more than 70,000 chemicals—are largely unknown to the public but are crucial for manufacturing processes. More than 1,000 new organic chemical products with commercial value are created each year. These chemicals are both useful and dangerous. Some of the causes of danger lie in the amounts generated and the methods used for disposal. Waste reduction at the source is the best solution.

Scanlon, Raymond D., ed. **Hazardous Materials, Hazardous Waste: Local Management Options.** Practical Management Series. Washington, DC: International City Management Association, 1987. 223p. ISBN 0-87326-052-X.

This book was written for local government managers, fire chiefs, elected officials, emergency planners, and anyone else responsible for programs to protect communities from hazardous material and toxic waste. Some of the dangers are pollution from hazardous products of local businesses, hazardous cargo spills on railroads and highways, landfills that cause contamination, and underground storage tanks that leak. The book discusses the complex problems regarding storage, transportation, and safety measures to be taken to prevent accidents and how to respond to emergencies. It also contains an excellent glossary and a list of sources of assistance. Additional references are also listed.

Technologies and Management Strategies for Hazardous Waste Control. Washington, DC: U.S. Congress, Office of Technology Assessment, 1983. 407p. No ISBN.

This book is the product of a three-year study made by the Office of Technology Assessment on hazardous waste management. The request for such a study was made by the House Committee on Energy and Commerce. About 275 million metric tons of hazardous waste under federal and state regulation are generated annually. Millions of tons disposed of in sanitary landfills pose substantial risks. Land disposal creates the risk of contaminating groundwater and can have adverse health effects. A key question is, Can risks and future cleanup costs be eliminated by limiting the use of land disposal and by suggesting attractive alternatives? Five policy options—continue with the current program, extend federal controls, establish federal fees on waste generators, classify wastes and waste management facilities as to their degree of hazardousness, and integrate federal environmental programs—are examined. The study stresses management strategies to protect human health and the environment.

Wynne, Brian. **Risk Management and Hazardous Waste: Implementation and the Dialectics of Credibility.** New York: Springer-Verlag, 1987. 447p. ISBN 0-387-18243-8.

This book is a printout of the project on Institutional Settings and Environmental Policies undertaken at the International Institute for Applied Systems Analysis of Vienna, Austria. Because this is a comparative book, the relative effectiveness of different regulatory practices is an implicit concern throughout. The argument in this book challenges convention by suggesting that the underlying approval of regulation in the developing climate of public justification reflects a misleading image of rationality and science. The central factor in this failure is the misconception that science is inherently uncertainty seeking; scientific approaches will therefore ensure a comprehensive appreciation of uncertainties. The eerie conclusion is that the public framework of regulations is driven by certain fundamental tenets of rational knowledge that no one believes in. The analytical framework throughout the volume is interpretative. This book presents a new basic orientation to management of hazardous wastes.

Toxic and Hazardous Waste Disposal

Chatterji, Manas, ed. **Hazardous Materials Disposal: Siting and Management.** Brookfield, VT: Gower Publishing Company, 1987. 331p. ISBN 0-566-05293-8.

This publication concentrates on the location and management of such facilities as nuclear power plants, nuclear waste disposal sites, hazardous material and toxic waste disposal sites, landfills, and incinerators, with emphasis on the not-in-my-backyard feeling. Communities need such facilities, but the question is where to install the unwelcome ones so they will pose no serious health problems. Case studies of Three Mile Island, Love Canal, Bhopal, and Chernobyl are discussed. A selected bibliography of location allocation is included.

Citizens Clearinghouse for Hazardous Wastes. **Deep Well Injection: An Explosive Issue.** Arlington, VA: The Clearinghouse, 1985. 73p. No ISBN.

This publication came out of a Citizens Clearinghouse for Hazardous Wastes (CCHW) roundtable meeting at which the main topic discussed was deep well disposal. Such disposal threatens the environment, water supplies, and people's health. Operation of deep wells is regulated by two federal laws: the Safe Drinking Water Act, which protects drinking water, and the Resource Conservation and Recovery Act, which regulates disposal by injection under a permit program. Deep well injection is the most frequently used method of hazardous waste disposal because it is cheap for industry to use and the EPA supports it. The CCHW roundtables are made possible by the support of the North Shore Unitarian/Universalist Program of New York.

Disposal of Industrial and Domestic Wastes: Land and Sea Alternatives. Board on Ocean Science and Policy, Commission on Physical Sciences,

Mathematics and Resources, National Research Council. Washington, DC: National Academy Press, 1984. 210p. ISBN 0-309-03484-1.

This study was conducted by 55 individuals drawn from industry, government, academia, and public interest groups. They were given the task of examining two specific cases: (1) the 106-Mile Ocean Waste Disposal Site off the New Jersey coast, the largest U.S. ocean site for disposal of industrial waste, and (2) the sewage sludge disposal problem in Los Angeles and Orange counties, California. The researchers evaluated criteria for land and ocean disposal of wastes as they made their study. The billions of tons of solid waste material produced annually must be managed by recycling, treatment, storage, or disposal; this situation has caused clashes among conflicting interests. Some waste material is harmless to humans and the environment, some can be made harmless through technological processes, and some must be handled carefully to prevent it from causing harm.

Hershkowitz, Allen. **Garbage Burning: Lessons from Europe: Consensus and Controversy in Four European States.** New York: INFORM, 1986. 53p. ISBN 0-918780-34-9.

The Nassau Neighborhood Network on Long Island, concerned about the hazards of landfills and garbage burning, and members of INFORM spent three weeks in Europe studying the methods of municipal waste disposal in Sweden, West Germany, Switzerland, and Norway. The researchers had three goals—to visit sites that rely heavily on resource recovery, to visit facilities that could offer the most useful insights on the burning of garbage, and finally to interview qualified people who had lengthy experience in overseeing resource recovery. The aim of this report is to discuss some of the issues associated with burning of garbage in Europe and how some of the information received might be used in the United States in materials separation, recycling, incineration, and landfills.

Kullenberg, Gunnar, ed. **The Role of the Oceans as a Waste Disposal Option.** NATO ASI Series C: Mathematical and Physical Sciences, Vol. 172. Published in cooperation with NATO Scientific Affairs Division. Boston: D. Reidel Publishing Company, 1985. 725p. ISBN 90-277-2209-9.

This volume contains the proceedings of the NATO Advanced Research Workshop on Scientific Basis for the Role of Oceans as a Waste Disposal Option held in Vilamoura Alsarve, Portugal, April 24–25, 1985. The papers deal with such subjects as oceans as a waste disposal option, sewage treatment and disposal, engineering of ocean outfalls, fresh water as a waste disposal system, sewage sludge disposal options, the North Sea ecosystem's behavior in relation to waste disposal, ecological and human health criteria, disposal of sewage in dispersive and nondispersive areas, metal pollution in the Great Lakes, and practice and assessment of sea dumping of radioactive waste.

Scholze, R. J., and others, eds. **Biotechnology for Degradation of Toxic Chemicals in Hazardous Wastes.** Park Ridge, NJ: Noyes Data Corporation, 1988. 697p. ISBN 0-8155-1148-5.

This book is a compilation of papers presented at the International Conference on Innovative Biological Treatment of Toxic Wastewaters organized by the Consortium for Biological Waste Treatment Research and Technology held in Arlington, Virginia, June 1986. Keynote addresses were given on biological treatment of toxics in wastewater and biological treatment in hazardous waste management, followed by papers dealing with the state of the art of biological treatment technology for toxic wastewater management.

Squires, Donald F. **The Ocean Dumping Quandary: Waste Disposal in the New York Bight.** Albany, NY: State University of New York Press, 1983. 226p. ISBN 0-87395-688-5.

This book was written to show the interaction between humans and the ocean environment since the ocean has become a repository for sewage sludges, industrial wastes, toxic substances, and spoils of dredging. It also shows how government and science try to cope with increasing pollution of the area. The author discusses the resources and people in the New York Bight region, legislation for improvement, and costs of dumping in the Bight to human health and the ecosystem, as well as the economic impact. The last chapter discusses the future of the Bight. Illustrations are found throughout the book. References are given for each chapter.

Subramanian, S. K., ed. **Treatment and Disposal of Hazardous Wastes from Industry: Some Experiences.** Tokyo: Asian Productivity Organization, 1983. 197p. ISBN 92-833-2005-0.

This volume is a report of the Symposium on Disposal and Recycling of Industrial Hazardous Wastes, held in Tokyo in 1982, and sponsored by the Asian Productivity Organization. Eleven countries were represented. Section 1 states the current status on disposal of hazardous waste in the countries participating. Section 2 presents papers delivered at the symposium, dealing with environmental and health effects of hazardous waste and regulations concerning treatment of hazardous waste. Bavaria had the highest environmental standards for the treatment and disposal of hazardous waste. The book ends with a list of recommendations made at the symposium as well as a list of participants, most of whom were from Asian countries. The book includes charts and maps used by authors of the papers.

Health

Freedberg, Louis. **America's Poisoned Playgrounds: Children and Toxic Chemicals.** Washington, DC: Conference on Alternative State and Local Policies, 1983. 54p. ISBN 0-89788-074-9.

Published jointly with *Youth News,* this study shows that many playgrounds are built in undesirable places, such as landfills that contain toxic chemicals and abandoned industrial plants that still have dangerous aspects even after the plants are demolished. Pesticides used in parks and playgrounds present health risks to children, and the toxic chemicals used to preserve wood equipment on playgrounds can be dangerous also. Children are thus exposed to a wide range of toxic substances. The book gives some recommendations to help lower the exposure to dangerous toxic substances that local governments should consider when choosing playground sites.

Nelkin, Dorothy, and Michael S. Brown. **Workers at Risk: Voices from the Workplace.** Chicago: University of Chicago Press, 1984. 220p. ISBN 0-226-57127-0.

This book is based on interviews with people who work with chemicals in various occupations, such as technicians in chemical and industrial plants, technical workers in production firms, laboratory technicians, fire fighters, and maintenance workers. The experiences of people who work with toxic chemicals, how they feel about working with substances that may affect their health, and how they cope with risks and the dangers they face are discussed in the book. Experiences of individual workers are cited. Appendix 1 gives biographical information on the workers who were interviewed. Appendix 2A gives the workers' perceptions of health effects from chemical exposure, and Appendix 2B lists major health effects of substances identified by respondents. Appendix 3 gives the respondents' toxicological concepts. The final appendix is a sample of the material safety data sheet used.

Weir, David. **The Bhopal Syndrome: Pesticides, Environment and Health.** San Francisco: Sierra Club Books, 1987. 210p. ISBN 0-87156-718-0.

The author discusses the tragic Bhopal incident and indicates that there are many unseen "slow-motion Bhopals" occurring from industrial pollution. He calls this the "Bhopal Syndrome," and argues that it must be stopped. The value systems of industrial enterprises must be changed so that health, safety of people, and the environment are of uppermost importance. Since the Bhopal pesticide disaster, other industrial accidents of major ecological significance have happened, such as the Chernobyl incident. Major industrial accidents over the past several decades and U.S. chemical accidents are listed in the appendix.

Pesticides

Barrons, Keith C. **Are Pesticides Really Necessary?** Chicago: Regnery Gateway, 1981. 245p. ISBN 0-89526-888-4.

After a survey of students, reporters, and others, the author believes that most people are unaware of the proven benefits of pesticides. He believes that the negative aspects of pesticides are too often emphasized. This book provides the reader with some insight into the benefits of pest control chemicals, which must be balanced against the risks that do exist. Part 1 of the book provides an examination of the many defenses—natural and human-contrived—that keep pests from overwhelming humankind. Part 2 describes a number of situations in which biological, genetic, and cultural controls have appeared adequate and where pesticides have been needed for health protection, and Part 3 deals with the question of pesticide safety, for people and for the environment.

Boardman, Robert. **Pesticides in World Agriculture: The Politics of International Regulation.** Macmillan International Political Economy Series. London: Macmillan, 1986. 221p. ISBN 0-333-37417-7.

This book is divided into three main sections. The first examines the broader context of international relations, where regulatory activity on pesticides takes place, and the main structures of the international pesticide economy. The second section deals with residues, registration, and use of chemicals and the environment. The last main section deals with issues of agricultural development and pesticides and the regulatory changes in developing countries. A final chapter deals with the politics of international regulation technology and regulatory functions and the pesticides regime.

Bogard, William. **The Bhopal Tragedy: Language, Logic, and Politics in the Production of a Hazard.** Boulder, CO: Westview Press, 1989. 154p. ISBN 0-8133-7786-2.

This book describes the events leading to the Bhopal tragedy. Many words have been used to describe the Bhopal incident—accident, tragedy, catastrophe, disaster, crisis, as well as sabotage, conspiracy, experiment, and massacre. No one knows how many people have died as a result of that event. As a sociologist, the author investigated the tragedy and its relevance to issues in the areas of social conflict and social organization, but found it extended to the political economy of development in the Third World and also to examinations of the hazards in our environment. He leaves us with two questions: How do we balance the risks and benefits of our increasingly hazardous technologies? and, How can we be sure there will be no more Bhopals? An extensive bibliography is included.

Bosso, Christopher J. **Pesticides and Politics: The Life Cycle of a Public Issue.** Pittsburgh, PA: University of Pittsburgh Press, 1987. 294p. ISBN 0-8229- 3547-3.

This study is about the development of policy in the control of pesticides. Emphasis is placed on political change. The individual chapters on such topics as the pesticides paradigm, the apotheosis of pesticides, changes in the 1960s, environmentalism and policy changes, the policy pendulum, the endless pesticides campaign, and the pesticides perspective are edited together into a broader documentary on policy transformation.

Carson, Rachel. **Silent Spring.** 25th anniversary ed. Boston: Houghton Mifflin, 1987 (first published 1962). 368p. ISBN 0-395-45390-9.

This book is concerned with the indiscriminate misuse of chemical insecticides and weed killers and how they affect our environment, damage wildlife, and become hazardous to humans. Carson argues that pesticides should be used safely and effectively, and that their use should be reduced to a minimum. She also promotes the search for alternative means to control bugs. The response of many scientists and chemists to this book has been that Carson exaggerates the picture. She provides heavy documentation to prove her points.

Cecil, Paul F. **Herbicidal Warfare: The RANCH HAND Project in Vietnam.** Praeger Special Studies—Scientific. New York: Praeger Publishers, 1986. 289p. ISBN 0-275-92007-0.

This book tells the story of the chemical warfare that took place in Vietnam, giving the mission and controversies surrounding it. The RANCH HAND Vietnam Association meets once a year to discuss what took place. One member is quoted as saying, "We have to tell lies, the truth is just too unbelievable." Operation RANCH HAND attacked the enemy's environment by destroying crops and forests. The book has detailed notes for each chapter. Appendix A provides a glossary of terms and abbreviations, and Appendix B gives a description of military herbicides. There is also a listing of participants consulted and documents and books used.

Cremlyn, R. J. **Pesticides: Preparation and Mode of Action.** New York: John Wiley & Sons, 1978. 240p. ISBN 0-471-99631-9.

This book discusses the growth in the application and sophistication of chemical pesticides, which has been particularly rapid since the 1940s. The physio-chemical factors and biochemical reactions important in pesticides are discussed as an introduction to the subsequent chapters dealing with the major chemical groups used to control different kinds of pests. The increasing awareness of the dangers of environmental pollution arising from the widespread use of chemical pesticides is reflected in the emphasis placed on safer and more selective chemicals. This volume is particularly valuable to individuals who want fundamental information on the use of pesticides.

Dover, Michael J. **A Better Mousetrap: Improving Pest Management for Agriculture.** Study 4. Washington, DC: World Resources Institute, 1985. 84p. ISBN 0-915825-09-0.

The author looks at pest control methods, chemicals, and technologies to see which offer the safest protection. The ways in which pest control methods are applied, and not the pest control methods themselves, decide whether they are safe or not. Once released into the environment, DDT and most chlorinated hydrocarbons can take years to break down; their toxic chemicals can accumulate in living organisms. Pesticide poisoning is a serious problem that can be prevented through better management. Less toxic chemicals or nonchemical techniques should be used in place of toxic ones. Indiscriminate use of pesticides causes environmental hazards. Integrated pest management (IPM) has both economic and environmental benefits, because it redefines rather than just refines pest control strategies. The volume contains several figures and tables.

Dover, Michael, and Brian Croft. **Getting Tough: Public Policy and the Management of Pesticide Resistance.** Study 1. Washington, DC: World Resources Institute, 1984. 92p. ISBN 0-915825-03-1.

Damage to the environment, threats to public health, and losses in crop production are serious problems associated with pesticide resistance discussed by the authors. The book tells how resistance management offers hope against immunity to pesticides and suggests 25 policy actions, such as stepped-up monitoring and mandatory and voluntary control of the use of pesticides.

Everest, Larry. **Behind the Poison Cloud: Union Carbide's Bhopal Massacre.** Chicago: Banner Press, 1986. 192p. ISBN 0-916650-26-X.

To obtain necessary information for this book the author spent six weeks in New Delhi, Bombay, and Bhopal, where he interviewed doctors caring for the victims of the Bhopal incident, government officials, and people who live near the Union Carbide Bhopal plant. He feels there has been a complete coverup by the U.S. chemical industry, the news media, and the U.S. government, and that there is a great deal that the U.S. public did not hear about the accident. The book has several pages of photos taken by author.

Gough, Michael. **Dioxin, Agent Orange: The Facts.** New York: Plenum Press, 1986. 290p. ISBN 0-306-42247-6.

Dioxin, an unseen, undetected substance lurking everywhere, is the most toxic chemical known. It is considered a "cancer-causing chemical," and its use in war constitutes chemical warfare. As a result of its use in Vietnam there have been many court cases. The author, a former official of the Technology Assessment Office of the U.S. Congress, reviews the health and environmental aspects of dioxin.

Holden, Patrick W. **Pesticides and Groundwater Quality: Issues and Problems in Four States.** Washington, DC: National Academy Press, 1986. ISBN 0-309-03676-3.

Holden wrote this book for the Board on Agriculture, National Research Council. The study analyzes the nature and scope of groundwater contamination by pesticides in California, New York, Wisconsin, and Florida, and how the problem of pesticide residues in groundwater affects public health and the environment. A list of individuals interviewed or consulted is given in the appendix. The book includes a bibliography.

Khan, Shahamat U. **Pesticides in the Soil Environment.** Fundamental Aspects of Pollution Control and Environmental Science, 5. New York: Elsevier/North-Holland Inc., 1980. 240p. ISBN 0-444-41873-3.

Pesticides, used extensively in the production of food, are viewed with suspicion and concern since they find their way into the soil and contaminate it. This book discusses pros and cons regarding the use of pesticides. The author begins by classifying the many pesticides into herbicides, insecticides, and fungicides. He discusses the behavior and fate of pesticides in the soil in terms of physiochemical and microbiological processes, the interactions between pesticides and the soil. The last chapter discusses the problem of minimizing pesticide residues in the soil, because to protect crops and control pests there will be a continuing need for the use of chemicals.

Morehouse, Ward, and M. Arun Subramaniam. **The Bhopal Tragedy: What Really Happened and What It Means for American Workers and Communities at Risk.** New York: Council on International and Public Affairs, 1986. 190p. ISBN 0-936876-47-6.

This is a preliminary report prepared for the Citizens Commission on Bhopal. It is based on a conference held March 20–21, 1985, on the theme "After Bhopal: Implications for Developed and Developing Nations," organized by the Workers' Policy Project in New York City. The book was written to tell what really happened on that night in December 1984, and to express the authors' feelings that we must not let this event be forgotten without making a determined effort to change the way in which we deal with hazardous industries and substances.

Repetto, Robert. **Paying the Price: Pesticide Subsidies in Developing Countries.** Research Report, no. 2. Washington, DC: World Resources Institute, 1985. 27p. ISBN 0-915825-12-0.

This small volume shows how pesticides are subsidized in developing countries and how the governments of those countries do not realize either the financial burden or the environmental and human costs of subsidization.

Seagrave, Sterling. **Yellow Rain: A Journey through the Terror of Chemical Warfare.** New York: M. Evans, 1981. 316p. ISBN 0-87131-349-9.

This book, written in a very easy-to-read style, tells the story of yellow rain. Explanatory notes are given for each chapter, along with a bibliography.

Weir, David, and Mark Schapiro. **Circle of Poison: Pesticides and People in a Hungry World.** San Francisco: Institute for Food and Development Policy, 1981. 100p. ISBN 0-935028-09-9.

The authors wrote this book after one of them, Weir, bought a package of Kool-Aid in Afghanistan and discovered it contained cyclamates—after they were banned in the United States. In subsequent study, they found that items banned in United States were being exported—for example, Tris-treated baby clothes. Banned pesticides are being exported and brought back into the United States as residues on products. The authors hope to help break the pesticide circle of poison that is affecting our environment. The book has a glossary; a list of chemical companies producing, buying, and selling hazardous pesticides in the Third World; and a list of pesticides used in foreign countries on food exported to the United States.

Westing, Arthur H. D., ed. **Herbicides in War: The Long-Term Ecological and Human Consequences.** Stockholm International Peace Research Institute (SIPRI). London: Taylor & Francis, 1984. 210p. ISBN 0-85066-265-6.

This book, prepared in cooperation with the U.N. Environment Programme, includes papers presented at the International Symposium on Herbicides and Defoliants in War: The Long-Term Effects on Man and Nature, held in Ho Chi Minh City, Vietnam, January, 13–20, 1983. Environmentalists' approach to the study of military activity is a different but valuable way of thinking. They look for ways to prevent world environmental damage. This is a work on the ecological and human consequences of the use of herbicides in war. The book has four appendices, including a bibliography.

Asbestos

Amaducci, Sandro, ed. **Asbestos: Directory of Unpublished Studies.** 2d ed. New York: Elsevier Science Publishing Co., 1986. 222p. ISBN 1-85166-073-9.

This second edition of the directory has been revised and enlarged to include technology and asbestos substitutes as well as health aspects. About 2,000 researchers and research centers around the world were surveyed to promote international cooperation. Studies are arranged by country and research center, giving stages of development, sponsors, and publications. Also included are a subject index, addresses of participating institutions, and names of research scientists. Health aspects are still stressed, as in the 1982 edition.

Benarde, Melvin A., ed. **Asbestos: The Hazardous Fiber.** Boca Raton, FL: CRC Press, 1990. 490p. ISBN 0-8493-6354-3.

Asbestos has always been a controversial mineral. Many authors who contributed to this volume are associated with Temple University's Asbestos Abatement Center and the U.S. EPA, Region III Satellite Center. The editor sets the stage for this book by writing on the history and state of the problem of asbestos. Adverse health effects in occupations and mere environmental exposure to asbestos are discussed in the book. Safety and protective equipment against asbestos and, finally, the disposal of asbestos are discussed. Three appendices give additional information about this hazardous fiber. References are found at the end of each chapter. Photos and tables are found in many chapters.

Hodgson, A. A., ed. **Alternatives to Asbestos: The Pros and Cons.** Critical Reports on Applied Chemistry, Vol. 26. Published for the Society of Chemical Industry. New York: John Wiley & Sons, 1989. 195p. ISBN 0-471-92353-2.

This volume is divided into three parts. Part I discusses the alternative materials that could be used in place of asbestos. Part II discusses the feasibility of substitution in regard to availability and cost. Part III discusses in detail the health effects of exposure to asbestos and other fibers. Each part has a list of references. Parts I and II have several tables.

Ouellette, Robert P., and others. **Asbestos Hazard Management: Guidebook to Abatement.** Lancaster, PA: Technomic Publishing Company, 1987. 399p. ISBN 0-87762-488-7.

Asbestos is no longer in great demand because of fear of its toxicity. However, the asbestos abatement industry has grown because of the need to remove asbestos from buildings. The book discusses use of asbestos in schools and public buildings and the hazards it poses, with laws and regulations controlling it. The book has a lengthy bibliography.

Peters, George A., and Barbara J. Peters. **Asbestos Review and Update: Supplement to Sourcebook on Asbestos Diseases, Medical, Legal, and Engineering Aspects.** New York: Garland Law Publishing, 1987. 150p. No ISBN.

This small volume updates the two volumes of the *Sourcebook on Asbestos Diseases*, published in 1980. It provides information regarding health, historical, technical, engineering, and legal aspects of asbestos. The book discusses the hazards connected with asbestos, and how they might be controlled. A detailed chapter is included on substitutes for asbestos. Chapters include lists of suggested readings, some figures, and tables. This is a valuable little book for anyone concerned about asbestos.

Schreier, H. **Asbestos in the Natural Environment.** Studies in Environmental Science, no. 37. New York: Elsevier Science Publishing Co., 1989. 159p. ISBN 0-444-88031-3.

Asbestos minerals have unusual properties—they can be woven, molded, and added to other materials to form superior products—that make them desirable for industrial use yet hazardous to human health. Asbestos materials have deleterious effects on soil and plant ecology. Researchers have focused attention on the physical and chemical properties of asbestos fibers to understand the environmental problems they cause. This book brings together the various aspects pertaining to asbestos in the natural environment—asbestos in the aquatic environment, asbestos in the soil environment, asbestos in plant growth, and other environmental concerns. There are 26 pages of bibliographical references and a list of tables and figures.

Skinner, H. Catherine W., Malcolm Ross, and Clifford Frondel. **Asbestos and Other Fibrous Materials.** New York: Oxford University Press, 1988. 204p. ISBN 0-19-503967-X.

Asbestos is a fibrous inorganic material mined and used because of its unusual chemical and physical properties. The hazards connected to asbestos are due mostly to the inorganic fibers; thus miners and people who manufacture products containing asbestos are affected. Many nonoccupational hazards are found in insulation in homes, schools, public buildings, brake linings of cars, hair driers, and many other products. This book is a study of fibrous inorganic materials and the health and biological effects associated with exposure to asbestos. The book has many tables and figures, along with several pages of bibliographical references. Three appendices list fibrous minerals, synthetic fibers, and brucite. A glossary is also very helpful.

Willis, George, and P. Reynolds. **Asbestos in the Urban Environment: A Manual of Control.** 2d ed. Kingston-upon-Thames (Surrey), England: Environmental Information and Analysis Publications, 1984. 156p. No ISBN.

This book begins by discussing the chemical properties of asbestos. It then delves into the hazards it poses to health and legislative actions taken to protect those exposed to asbestos. "Stripping," the method for removal of asbestos, and alternatives to asbestos in the building industry are discussed.

Laws and Regulations

Bierlein, Lawrence W. **Red Book on Transportation of Hazardous Materials.** 2d ed. New York: Van Nostrand Reinhold, 1988. 1,203p. ISBN 0-442-21044-2.

This book gives the laws and regulations pertaining to the transportation of hazardous materials.

DiMento, Joseph F. **Environmental Law and American Business: Dilemmas of Compliance.** Environment, Development, and Public Policy. Environment Policy and Planning Series. New York: Plenum Press, 1986. 228p. ISBN 0-306-42168-2.

This volume deals with environmental law and indicates that numerous violations are committed even though the law is enforced with extensive expenditures and many people are sympathetic to environmental objectives. Many violations involve American business; examples given include an industrialist who made a fortune by illegally dumping dangerous chemicals, a businessman who allowed spillover of a pollutant into a protected stream, and a hazardous waste site in California that became an inferno—an incident for which no one would assume responsibility. The aim of this book is to suggest policy reforms to induce businesses to comply with environmental law. A list of cases appears in the appendix. There is an extensive bibliography.

Jones, David, and Jeffrey Tryens, eds. **Legislative Sourcebook on Toxics.** Washington, DC: National Center for Policy Alternatives, 1986. 235p. ISBN 0-89799-092-7.

This book was written for state legislators, governors, administrators, civic activists, and those concerned about toxic problems. The book lists 40 bills and laws from 22 states, with short descriptions, sponsors, and further references. Some of the topics discussed include abandoned and illegal dumps and facilities, consumer protection, hazardous waste and material regulation, liability and compensation, pesticides, and small-quantity generators.

McDowell, George B., ed. **Pesticide Guide: Registration, Classification and Applications.** Neenah, WI: J. J. Keller and Associates, 1979. 1 vol. (various paging). ISBN 0-934674-12-4.

This book explains federal regulations concerning pesticide packaging, labeling, storage, disposal, and the like in a section headed "Compliance." The second section, "Regulations," records all federal regulations dealing with pesticides, and the third section, "Reference," discusses forms required by the government for reporting use of pesticides, enforcement of pesticide regulations, and pending state and federal legislation. This is a loose-leaf publication, and thus can be updated easily.

Mallow, Alex. **Hazardous Waste Regulations: An Interpretive Guide.** New York: Van Nostrand Reinhold, 1981. 403p. ISBN 0-442-21935-0.

The purpose of this book is to make the complex laws and regulations regarding hazardous waste easy to understand. The book has a list of hazardous wastes, standards for treatment, storage and disposal facilities, and transportation guidelines. There are five appendices defining words and phrases that appear in the regulations, hazardous waste regulations, proposed hazardous waste regulations, and Department of Transportation regulations giving contents only. One appendix gives a complete copy of the Resource Conservation and Recovery Act.

Mendeloff, John M. **The Dilemma of Toxic Substance Regulation: How Overregulation Causes Underregulation at OSHA.** MIT Press Series on Regulation of Economic Activity, no. 17. Cambridge, MA: MIT Press, 1988. 321p. ISBN 0-262-13230-3.

This book deals with government regulations to reduce the workshop risks caused by toxic substances. The author shows that overregulation can cause underregulation. He also shows how this situation arose and what can be done to alleviate it. The book has eight appendices and extensive notes on each chapter.

Moskowitz, Joel S. **Environmental Liability and Real Property Transactions: Law and Practice.** Real Estate Practice Library. New York: John Wiley & Sons, 1989. 384p. ISBN 0-471-61390-8.

This book was written by a lawyer for those involved in real estate transactions, giving advice on how to manage the environmental aspects of transactions and discovering even small signs of environmental trouble. One chapter is devoted to New Jersey's Environmental Cleanup Responsibility Act. Another chapter discusses the cleanup of such materials as asbestos, PCBs, chlorinated hydrocarbons, petroleum products, pesticides, and heavy metals.

Musselman, Victoria Cooper. **Emergency Planning and Community Right-To-Know: An Implementer's Guide to SARA Title III.** New York: Van Nostrand Reinhold, 1989. 204p. ISBN 0-442-20555-4.

Planning prevention and understanding the nature of potential hazards from handling toxic waste and hazardous material are two requirements stressed in the Emergency Planning and Community Right-To-Know Law. The entire book is devoted to the law and gives a complete picture. Those readers interested in a specific aspect of the law should consult the table of contents.

Peters, George A., and Barbara J. Peters. **Sourcebook on Asbestos Diseases: Medical, Legal, and Engineering Aspects.** New York: Garland Publishing Inc., 1980–1986. Vol. 1 (various paging). ISBN 0-8240-7175-1. Vol 2, 843p. ISBN 0-8240-7299-5.

These two volumes, written by attorneys, deal with diseases resulting from the inhalation and ingestion of asbestos fibers. The legal and medical aspects of asbestos diseases are discussed and ways to protect the worker are given. The authors review the case law and workers' health and safety standards in ten countries. Warnings required by the Occupational Safety and Health Administration are provided. Both volumes have bibliographies from the United States, South Africa, and Western Europe covering the period 1899–1982. The books have author, title, and subject indexes. Lists of asbestos producers, manufacturers, and trade products with trade names are given. An interesting section gives the text of five most frequently cited court decisions regarding liability for asbestos-related diseases. Specifications for asbestos abatement are included. Those concerned with asbestos abatement as well as liability insurers will find the book useful.

Reese, Crain E. **Deregulation and Environmental Quality: The Use of Tax Policy To Control Pollution in North America and Western Europe.** Westport, CT: Quorum Books, 1983. 495p. ISBN 0-89930-018-9.

The author discusses the regulation of polluters as a solution to the environmental crisis. The most cost-effective method for controlling the production, use, and disposal of toxic substances, the disposal of hazardous waste, and the promotion of health and safety in the working environment would be the regulation and reduced production of solid waste. The book studies the situation in Canada, France, West Germany, Sweden, the United Kingdom, and the United States. Comparisons are made among direct grant subsidization, tax-incentive subsidization, and pollution taxation tactics used in the six countries. Many figures and tables are used in the book. Nine appendices explain such topics as EPA guidelines, EPA regulations, environmental excise taxes, and a list of state solid waste agencies. There is a very good chapter on the environmental excise taxes on U.S. crude oil, chemicals, and hazardous materials.

The Toxic Substances Control Act: Overview and Evaluation. Policy Research Project Report, no. 50. Austin: Lyndon B. Johnson School of Public Affairs, University of Texas, 1982. 234p. ISBN 0-89940-650-5.

This volume is a report by the Toxic Substances Control Act Policy Research Project, which is composed of graduate students and faculty. The purpose of the report is to show the role the United States played in drafting and implementing the act. The legislative history of TSCA is given, along with health and environmental risks and economic impacts of controlling toxic substances. The volume ends by discussing management and implementation of the act. This report was prepared for the Economic Council of Canada.

Wagner, Travis P. **The Hazardous Waste Q & A: An In-Depth Guide to the Resource Conservation and Recovery Act and the Hazardous Materials Transportation Act.** New York: Van Nostrand Reinhold, 1990. 404p. ISBN 0-442-23842-8.

This volume is written in a question-and-answer format, making it easy to understand. It deals with the regulation of hazardous waste under the Resource Conservation and Recovery Act and the Hazardous Materials Transportation Act. The book is of value to anyone interested in hazardous waste regulations. There are seven appendices to help the reader better understand the book. Appendix A is especially important, because it defines terms used in the hazardous waste regulations.

Worobec, Mary Devine, and Girard Ordway, eds. **Toxic Substances Controls Guide: Federal Regulation of Chemicals in the Environment.** Washington, DC: BNA Books, 1989. 239p. ISBN 0-87179-632-5.

This book explains major laws affecting the chemical life cycle and statutes dealing with chemicals in the environment and workplace, and the transportation and disposal of such materials. It explains government regulation of the chemical industry in an easy-to-understand style for those with a chemistry background.

Articles and Government Documents

Toxic and Hazardous Waste

General

Ashbrook, P. C., and P. A. Reinhardt. "Hazardous Wastes in Academia." *Environmental Science and Technology* 19 (December 1985): 1150–1155.

Basta, N., and others. "What Are Your Views on Hazardous Wastes?" *Chemical Engineering* 92 (March 4, 1985): 58–66.

Brinkman, D. W. "Used Oil: Resource or Pollutant?" *Technology Review* 88 (July 1985): 46–51.

Cheremisinoff, P. N. "High Hazard Pollutants: Asbestos, PCBs, Dioxins, Biomedical Wastes." *Pollution Engineering* 21 (February 1989): 58–65.

Coates, Vary T., and others. "Toxics '95: Outlook of Factors and Trends for Toxic Chemicals." *Toxic Substances Journal* 6 (Summer 1984): 26–43.

Cole, Henry S. "Toxic Chemical Information Systems and Right-To-Know." *Journal of Public Health Policy* 7 (Spring 1986): 28–36.

Davis, Charles E., and James P. Lester. "Hazardous Waste Politics and Policy: A Symposium." *Policy Studies Journal* 14 (September 1985): 47–168.

Eckhardt, Robert C. "The Unfinished Business of Hazardous Waste Control." *Baylor Law Review* 33 (Spring 1981): 253–265.

Florini, Karen L. "Issues of Federalism in Hazardous Waste Control: Cooperation or Confusion?" *Harvard Environmental Law Review* 6:2 (1982): 307–337.

Florini, Karen L., and others. "EPA's Delisting Program for Hazardous Wastes: Current Limitations and Future Directions." *Environmental Law Reporter* 19 (December 1989): 10558–10568.

Hirschhorn, J. S. "Cutting Production of Hazardous Waste." *Technology Review* 91 (April 1988): 52–61.

Krauss, Celene. "Grass Root Consumer Protests and Toxic Wastes: Developing a Critical Political View." *Community Developing Journal* 23 (October 1988): 258–265.

Lehr, Jay H. "The Future of the Hazardous Waste Program." *Environmental Forum* 4 (March 1986): 16–23.

McElfish, James M., Jr. "State Hazardous Waste Crimes." *Environmental Law Reporter* 17 (December 1977): 10465–10477.

Mazur, Allan. "The Journalists and Technology: Reporting about Love Canal and Three Mile Island." *Minerva* 22 (Spring 1984): 45–66.

Minges, J., and J. Graumann. "The Hazwaste Response." *American City and County* 101 (October 1986): 58–65.

Plaut, Jon. "Hazardous Waste Control and Industry." *Toxic Substances Journal* 5 (Spring 1984): 251–260.

Pugel, Thomas A., and Ingo Walter. "Toxic Substances and TNC Involvement in the Chemical and Pharmaceutical Industries." *Toxic Substances Journal* 4 (Summer 1982): 55–76.

Schneider, Claudine. "Hazardous Waste: The Bottom Line Is Prevention: Decades of Lip Service to the Advantage of Reducing Waste at the Source Have Yielded Little Progress; Stronger Federal Direction Is Needed." *Issues in Science and Technology* 4 (Summer 1988): 75–80.

Skinner, J. H., and N. J. Bassin. "The Environmental Protection Agency's Hazardous Waste Research and Development Programs." *Journal of the Air Pollution Control Association* 38 (April 1988): 377–387.

"Troubling Times with Toxics." *National Wildlife* 24 (February–March 1986): 26–36.

U.S. Congress. Senate. Committee on Environment and Public Works. Subcommittee on Toxic Substances, Environmental Oversight, Research, and Development. *Issues Related to the Use of, and Exposure to, Various Chemicals: Hearings, March 6 and July 5, 1989.* 101st Cong., 1st sess. Washington, DC: GPO, 1989. 409p.

Want, William L. "Hazardous Waste: A Business Primer; or, What Corporate America Must Know about This Enormous Problem; To Ignore It Might Lead to Bankruptcy Court." *Business and Economic Review [University of Southern California]* 34 (July–September 1988): 3–8.

Waxman, Henry A. "Chemicals—Looking for the Panacea in Pandora's Box." *World Affairs Journal* 5 (Winter 1986): 1–6.

Wu, J. S., and H. Hilger. "Evaluation of EPA's Hazard Ranking System." *Journal of Environmental Engineering* 110 (August 1984): 797–807.

Accidents

GENERAL

Harris, Elisa D. "Sverdlovsk and Yellow Rain: Two Cases of Soviet Noncompliance." *International Security* 11 (Spring 1987): 41–95.

Miller, Leonard A., and Robert S. Taylor. "The Enemy Below: EPA Plans Action on Leaking Underground Storage Tanks." *Environmental Law Reporter* 15 (May 1985): 10135–10143.

Neely, W. B., and R. W. Lutz. "Estimating Exposure from a Chemical Spilled into a River." *Journal of Hazardous Materials* 10 (February 1985): 33–41.

Scoville, W., and others. "Response and Cleanup Efforts Associated with the White Phosphorus Release, Miamisburg, Ohio." *Journal of Hazardous Materials* 21 (January 1989): 47–64.

Van Aerde, M., and others. "Estimating the Impacts of L.P.G. Spills during Transportation Accidents." *Journal of Hazardous Materials* 20 (December 1988): 375–392.

BHOPAL

Agarwal, A. "The Cloud over Bhopal." *New Scientist* 108 (November 16, 1985): 38–41.

Bowonder, B., and T. Miyake. "Managing Hazardous Facilities: Lessons from the Bhopal Accident." *Journal of Hazardous Materials* 19 (November 1988): 237–269.

Bryan, J. "What Future for Bhopal's Victims?" *New Scientist* 104 (December 24, 1984): 20–27.

Heylin, M. "Bhopal [special issue]." *Chemical and Engineering News* 63 (February 11, 1985): 14–65.

Kamlet, Kenneth. "High-Priority Pollution Control Issues in the Aftermath of Bhopal." *Toxic Substances Journal* 7:1–4 (1985/1986): 109–118.

———. "Industrial Pollution Control in the Aftermath of Bhopal." *Toxic Substances Journal* 6 (Summer 1984): 13–25.

Leonard, Richard. "After Bhopal: Multinationals and the Management of Hazardous Products and Processes." *Multinational Business* no. 2 (1986): 1–9.

Lepkowski, W. "Bhopal: Indian City Begins To Heal but Conflicts Remain." *Chemical and Engineering News* 63 (December 2, 1985): 18–32.

————. "Chemical Reaction: Safety after Bhopal: Will the Indian Tragedy Spur Increased Safety?" *Business and Society Review* (Summer 1986): 38–43.

Mahon, John F., and Patricia C. Kelley. "Managing Toxic Wastes: After Bhopal and Sandoz." *Long Range Planning* 20 (August 1987): 50–59.

Montgomery, Christian. "Reducing the Risk of Chemical Accidents: The Post-Bhopal Era." *Environmental Law Reporter* 16 (October 1986): 10300–10305.

"Mystery of Who Designed Bhopal's Plant." *New Scientist* 104 (December 7, 1984): 20–27.

"United States: District Court for the Southern District of New York Opinion in Re: Union Carbide Corporation Gas Plant Disaster at Bhopal, India in December, 1984." *International Legal Materials* 25 (July 1986): 771–802.

Varma, Vijaya Shankar, and Shiv Visvanathan. "Bhopal: Facts and Reflections." *Alternatives [Center for the Study of Developing Societies]* 11 (January 1986): 133–165.

"What Is Appropriate Public Policy Response to Bhopal?" *Environmental Forum* 4 (August 1985): 323–335.

LIABILITY

"CERCLA Litigation Update: The Emerging Law of Generator Liability." *Environmental Law Reporter* 14 (June 1984): 10224–10236.

Hadden, Susan G. "Labeling of Chemicals To Reduce Risk." *Law and Contemporary Problems* 46 (Summer 1983): 235–266.

Katzman, Martin C. "Chemical Catastrophes and the Courts." *Public Interest (New York)* (Winter 1986): 91–105.

Schroeder, Christopher, ed. "Federal Regulation of the Chemical Industry." *Law and Contemporary Problems* 46 (Summer 1983): 1–305.

RHINE RIVER

Capel, P. D., and others. "Accidental Input of Pesticides into the Rhine River." *Environmental Science and Technology* 22 (September 1988): 992–997.

Mossman, D. J., and others. "Predicting the Effects of a Pesticide Release to the Rhine River." *Journal of Water Pollution Control Federation* 60 (October 1988): 1806–1812.

Emergency Preparedness

Ajamie, Thomas R. "Emergency Planning for Hazardous Chemical Accidents: Elements of a Legislative Solution." *Journal of Legislation* 12 (Summer 1985): 195–212.

American Automobile Association. Foundation for Traffic Safety. *Local Response to Hazardous Materials Incidents and Accidents.* Falls Church, VA: 1986. 47p.

Bracken, Marilyn C. "Information Technology in Emergency Response and Hazardous Waste Management." *Information Society* 3:4 (1985): 361–369.

Burcat, Joel R., and Arthur K. Hoffman. "The Emergency Planning and Community Right-To-Know Act of 1986: An Explanation of Title III of SARA." *Environmental Law Reporter* 18 (January 1988): 10007–10027.

Dietz, A. G., Jr. "Liquefied Gaseous Fuels (LGF) Spill Test Facility Program." *Journal of Environmental Science* 28 (September–October 1985): 34–39.

Froebe, L. R. "State and National Resources for Community Spill Disaster Preparedness in the United States." *Journal of Hazardous Materials* 10 (February 1985): 107–124.

Gephart, Robert P., Jr. "Organization Design for Hazardous Chemical Accidents." *Columbia Journal of World Business* 22 (Spring 1987): 51–58.

Granito, John A. "Hazardous Materials Incidents: Improving Community Responses." *Management Information Service Report* 16 (January 1984): 1–12.

Heard, D. B. "Fail-Safe Devices for the Prevention of Hazardous Materials Spills." *Journal of Hazardous Materials* 13 (April 1986): 233–238.

Johnson, James H., Jr., and Donald J. Zeigler. "Evacuation Planning for Technological Hazards: An Emerging Imperative." *Cities* 3 (May 1986): 148–156.

"Local Response to Hazardous Materials Incidents and Accidents." *Transportation Quarterly* 40 (October 1986): 461–482.

Reuber, C., and others. "A Public Information and Notification System for Emergency Mitigation." *IEEE Transactions on Power Delivery* 3 (October 1988): 1356–1361.

Stull, J. O. "Protective Suits for Chemical Spill Response." *Chemical Engineering Progress* 83 (November 1987): 34–39.

Taylor, Lynda. "Title III: Bhopal's Baby." *Workbook [Southwest Research and Information Center]* 12 (October–December 1987): 132–137.

U.S. Congress. Senate. Committee on Environment and Public Works. *The Ability To Respond to Toxic Chemical Emergencies: Hearing, February 18, 1985.* 99th Cong., 1st sess. Washington, DC: GPO, 1985. 212p.

Environmental Aspects

Cannon, J. B., and others. "Environmental Effects of Fusion Power Plants: Effluents Other than Tritium." *Fusion Technology* 12 (November 1987): 341–353.

Coates, Ruth H. "Solvent Recovery Protects the Environment and Conserves Resources." *Industrial Wastes* 29:4 (July–August 1983): 16–21.

Eisenberg, D. M., and others. "Groundwater Protection in San Francisco Bay Area." *Journal of Environmental Engineering* 111 (August 1985): 431–440.

Fiksel, J. "Quantitative Risk Analysis for Toxic Chemicals in the Environment." *Journal of Hazardous Materials* 10 (July 1985): 227–240.

"Management of Toxins in the Environment." *American Journal of Agricultural Economics* 71 (December 1989): 1286–1304.

"Managing Your Environmental Audit." *Chemical Engineering* 92 (June 24, 1985): 37–43.

Persson, P. E. "Uptake and Release of Environmentally Occurring Odorous Compounds by Fish: A Review." *Water Research* 18:10 (1984): 1263–1271.

Roberts, P. J. W., and G. Toms. "Ocean Outfall System for Dense and Buoyant Effluents." *Journal of Environmental Engineering* 114 (October 1988): 1175–1191.

Schroeder, W. H., and D. A. Lane. "The Fate of Toxic Airborne Pollutants." *Environmental Science and Technology* 22 (March 1988): 240–246.

Smith, Turner T., Jr. "Environmental Damage Liability Insurance: A Primer." *Business Lawyer* 39 (November 1983): 333–354.

Swanson, Larry D. "Shifting the Burden of Environmental Protection [hazardous waste and toxic substances management policies]." *Journal of Economic Issues* 18 (March 1984): 251–274.

Thompson, Roger. "Preventing Groundwater Contamination." *Editorial Research Reports* (July 12, 1985): 519–536.

"Toxics and the Environment—An Issue Summarizing Process Design Information for Removal Methods of Phosphorus in Domestic Wastewaters." *Toxic Substances Journal* 8:2–3 (1988): entire issue.

U.S. Congress. House. Committee on Energy and Commerce. Subcommittee on Commerce, Transportation, and Tourism. *Control of Carcinogens in the Environment: Hearing, March 17, 1983.* 98th Cong., 1st sess. Washington, DC: GPO, 1983. 557p.

U.S. Congress. House. Committee on Energy and Commerce. Subcommittee on Oversight and Investigations. *Ground Water Monitoring: Hearing, April 29, 1985.* 99th Cong., 1st sess. Washington, DC: GPO, 1985. 194p.

U.S. Congress. House. Committee on Public Works and Transportation. Subcommittee on Investigations and Oversight. *Diffuse Toxic Pollutants in the Great Lakes Ecosystem: Hearing, April 14, 1988.* 100th Cong., 2d sess. Washington, DC: GPO, 1988. 108p.

———. *Hazardous Waste Contamination of Water Resources (Access to EPA Records): Hearing, December 2, 1982.* 97th Cong., 2d sess. Washington, DC: GPO, 1983. 100p.

———. *Hazardous Waste Contamination of Water Resources: Hearings, March 10–July 9, 1982.* 97th Cong., 2d sess. Washington, DC: GPO, 1983. 368p.

U.S. Congress. House. Committee on Public Works and Transportation. Subcommittee on Water Resources. *Toxic Pollution in the Great Lakes: Hearing, March 2, 1978.* 100th Cong., 2d sess. Washington, DC: GPO, 1988. 145p.

U.S. Congress. Senate. Committee on Environment and Public Works. *Groundwater Contamination by Toxic Substances: A Digest of Reports; A Report, November 1983.* 98th Cong., 1st sess. Washington, DC: GPO, 1983. 75p.

U.S. Department of the Army. Program Manager for Chemical Demilitarization. *Chemical Stockpile Disposal Program: Final Programmatic Environmental Impact Statement.* Aberdeen, MD: Aberdeen Proving Ground, 1988. 3 vols.

Export-Import Trade

Agege, Charles O. "Dumping of Dangerous American Products Overseas: Should Congress Sit and Watch?" *Journal of World Trade Law* 19 (July–August 1985): 403–410.

Applegate, Howard G., and C. Richard Bath. "Hazardous and Toxic Substances in U.S.-Mexico Relations [particularly Mexican use of pesticides banned in the United States]." *Texas Business Review* 57 (September–October 1983): 229–234.

Chetley, Andrew. "Not Good Enough for Us but Fit for Them: An Examination of the Chemical and Pharmaceutical Export Trades." *Journal of Consumer Policy* 9 (June 1986): 155–180.

Cohen, Susan L. "Exports of Hazardous Products from the United States: An Analysis of Consumer Product Safety Commission Policy." *George Washington Journal of International Law and Economics* 19:1 (1985): 123–163.

Harland, David. "Legal Aspects of the Export of Hazardous Products." *Journal of Consumer Policy* 8 (September 1985): 209–238.

Mokhiber, Russell, ed. "Special Issue: Export of Hazards." *Multinational Monitor* 5 (September 1984): 3–22.

Seferovich, Patrick B. "United States Export of Banned Products: Legal and Moral Implications [export to developing countries of domestically banned substances]." *Denver Journal of International Law and Policy* 10 (Spring 1981): 537–560.

Shaikh, Rashid A. "The Dilemmas of Advanced Technology for the Third World: Developed Nations Must Disclose the Hazards of Their Exports, But Poor Nations Also Have To Rethink Their Development Goals." *Technology Review* 89 (April 1986): 56–64.

Shuman, Eric. "Potentially Hazardous Merchandise: Domestic and International Mechanisms for Consumer Protection." *Vanderbilt Journal of Transnational Law* 16 (Winter 1983): 179–229.

Singh, Jang B., and V. C. Lakhan. "Business Ethics and the International Trade in Hazardous Wastes." *Journal of Business Ethics* 8 (November 1989): 889–899.

U.S. Environmental Protection Agency. Office of Toxic Substances. *Toxic Substances Control Act: A Guide for Chemical Importers/Exporters*. Washington, DC: GPO, 1984. 2 vols.

Health Effects

Andrade, Vibiana M. "The Toxic Workplace: Title VII [of the Civil Rights Act of 1964] Protection for the Potentially Pregnant Person." *Harvard Women's Law Journal* 4 (Spring 1981): 71–103.

Brown, Stephen L., and Benjamin E. Suta. "A General Overview of Approaches toward Comprehensive Exposure Assessments." *Toxic Substances Journal* 4 (Summer 1982): 23–27.

Buss, Emily. "Getting beyond Discrimination: A Regulatory Solution to the Problem of Fetal Hazards in the Workplace." *Yale Law Journal* 95 (January 1985): 577–598.

Carno, Tina. "A New Cause of Action for Massive Medical Toxic Injury." *Glendale Law Review* 6 (1984): 15–29.

"Controlling Chemical Contamination in the Workplace." *Harvard Environmental Law Review* 9:2 (1985): 249–398.

Corn, Jacqueline Karnell. "Vinyl Chloride: Setting a Workplace Standard: An Historical Perspective on Assessing Risk." *Journal of Public Health Policy* 5 (December 1984): 497–512.

Ferguson, J. S., and W. F. Martin. "An Overview of Occupational Safety and Health Guidelines for Superfund Sites." *American Industrial Hygiene Association Journal* 46 (April 1985): 175–180.

Freed, Cheryl. "The Tussock Moth Incident and the DDT Ban: Another Look." *Toxic Substances Journal* 5 (Autumn 1983): 88–98.

Hattis, Dale, and David Kennedy. "Assessing Risks from Health Hazards: An Imperfect Science." *Technology Review* 89 (May–June 1986): 60–67.

Johnson, Barry L. "Health Effects of Hazardous Waste: The Expanding Functions of the Agency for Toxic Substances and Disease Registry." *Environmental Law Reporter* 18 (April 1988): 10132–10138.

Miller, James C., III. "Comparative Data on Life-Threatening Risks." *Toxic Substances Journal* 5 (Summer 1983): 3–14.

———. "Occupational Exposure to Acrylonitriles: A Benefit/Cost Analysis." *Toxic Substances Journal* 4 (Winter 1982/1983): 223–233.

New Jersey. Drinking Water Quality Institute. *Maximum Contaminant Level Recommendations for Hazardous Contaminants in Drinking Water.* Trenton, NJ: 1987. 52p.

Onishi, Y., and others. "Computer-Based Environmental Exposure and Risk Assessment Methodology for Hazardous Materials." *Journal of Hazardous Materials* 10 (July 1985): 389–417.

Page, Norbert P. "Testing for Health and Environmental Effects: The OECD Guidelines." *Toxic Substances Journal* 4 (Autumn 1982): 135–153.

Paskal, Steven S. "Dilemma: Save the Fetus or Sue the Employer." *Labor Law Journal* 39 (June 1988): 323–341.

Rall, David P. "Toxic Agent and Radiation Control: Meeting the 1990 [U.S. Public Health Service's] Objectives for the Nation." *Public Health Reports* 99 (November–December 1984): 532–538.

Randall, Donna M. "Fetal Protection Policies: A Threat to Employee Rights?" *Employee Responsibilities and Rights Journal* 1 (June 1988): 121–128.

————. "Women in Toxic Work Environments: A Case Study and Examination of Policy Impact." In *Women and Work, 1985*, pp. 259–281. Edited by Laurie Larwood and others. Beverly Hills, CA: Sage Publications, 1985.

Rodricks, Joseph V. "Can Risk Assessment Be Improved?" *Toxic Substances Journal* 6 (Spring–Summer 1985): 178–185.

Stellman, Jeanne M., and Leslie R. Andrews. "The Assessment of Toxic Exposure in the Workplace." *Toxic Substances Journal* 4 (Autumn 1982): 104–115.

Timko, Patricia A. "Exploring the Limits of Legal Duty: A Union's Responsibilities with Respect to Fetal Protection Policies." *Harvard Journal of Legislation* 23 (Winter 1986): 159–210.

U.S. Congress. House. Committee on Government Operations. *Occupational Health Hazards: Joint Hearing, April 16, 1986, before the Intergovernmental Relations and Human Resources Subcommittee and the Employment and Housing Subcommittee.* 99th Cong., 2d sess. Washington, DC: GPO, 1986. 347p.

U.S. Congress. House. Committee on Science and Technology. Subcommittee on Investigations and Oversight. *Compensation for Exposure to Hazardous Substances: Hearing, August 12, 1982.* 97th Cong., 2d sess. Washington, DC: GPO, 1983. 487p.

————. *Neurotoxins at Home and in the Workplace: Hearings, October 8–9, 1985.* 99th Cong., 1st sess. Washington, DC: GPO, 1986. 285p.

————. *Relation of Exposure to Toxic Chemicals and Reproductive Impairment: Hearing, July 27, 1982.* 97th Cong., 2d sess. Washington, DC: GPO, 1983. 145p.

Wagstaff, D. Jesse. "Assessment of Human Exposure to Toxic Substances in Food." *Toxic Substances Journal* 4 (Winter 1982/1983): 184–198.

Wallace, Lance. "Measuring Direct Individual Exposure to Toxic Substances." *Toxic Substances Journal* 4 (Winter 1982/1983): 174–183.

Waller, Robin, and Sarah Rosenblatt. "Toxic Hazards in the Office Environment: Compensability of Claims Arising from the Office Equipment." *Toxic Substances Journal* 5 (Winter 1983/1984): 182–198.

Wang, Charleston C. K. "Toxic Agents, Carcinogens, Worker Health, Cost-Benefit Analysis and the Clamor for Reasonable OSHA [U.S. Occupational Safety and Health Administration] Regulations: A Survey of Judicial and Other Answers to a Complex Socio-Technological Controversy." *Northern Kentucky Law Review* 8:3 (1981): 589–629.

Wilkins, John R., III. "Exposure Assessment in Studies of Environmental Hazards: An Epidemiologic Perspective." *Toxic Substances Journal* 5 (Autumn 1983): 75–87.

Wolf, K. "Hazardous Substances and Cancer Incidence: Introduction to the Special Issue on Risk Assessment and Risk Management." *Journal of Hazardous Materials* 10 (July 1985): 167–178.

Laws and Regulations

GENERAL

Alexander, Beverly Z. "CERCLA 1980–1985: A Research Guide." *Ecology Law Quarterly* 13:2 (1986): 311–359.

Allen, Leslie. "Who Should Control Hazardous Waste on Native American Lands? Looking beyond *Washington Department of Ecology v. EPA*." *Ecology Law Quarterly* 14:1 (1987): 69–116.

Arthur, Jack L., and Roger L. Garrett. "Exposure Assessment of New Chemicals under Section 5 of the Toxic Substances Control Act (TSCA)." *Toxic Substances Journal* 4 (Autumn 1982): 80–103.

Arup, Christopher. "Chemical Notification Laws in the OCED Member Countries." *Journal of World Trade Law* 21 (February 1987): 47–66.

Ashford, Nicholas A. "Advisory Committees in OSHA and EPA: Their Use in Regulatory Decisionmaking." *Science, Technology and Human Values* 9 (Winter 1984): 72–82.

Balliveau, R. E., Fred Hoerger, and Richard Hinds. "Scientific Peer Review and the Regulation of Chemical Hazards." *Toxic Substances Journal* 6 (Autumn–Winter 1984): 78–96.

Black, Edward G. "California's Community Right-To-Know." *Ecology Law Quarterly* 16:4 (1989): 1021–1064.

Bowman, Ann O'M. "Intergovernmental and Intersectoral Tensions in Environmental Policy Implementation: The Case of Hazardous Waste." *Policy Studies Review* 4 (November 1984): 230–244.

"Burdens of Environmental Regulation on Private Property Ownership and Business Transactions: Reasonable or Unreasonable?" *Environmental Law Reporter* 18 (September 1988): 10348–10386.

Cannon, J. A. "The Regulation of Toxic Air Pollutants." *Journal of the Air Pollution Control Association* 36 (May 1986): 562–573; Discussion, 36 (September 1986): 986–996.

"CERCLA Symposium." *Washington University Journal of Urban and Contemporary Law* 31 (Winter 1987): 241–337.

Chess, Caron. "Looking behind the Factory Gates: Right-To-Know Laws Requiring Businesses To Divulge Information about Chemicals They Use May Help Improve Public Health." *Technology Review* 89 (August–September 1986): 42–47.

Cohen, Douglas A., and others. "Eroding the Doctrine: 'As Is' and Caveat Emptor under CERCLA." *Environmental Claims Journal* 2 (Autumn 1989): 29–41.

Connolly, Diane M. "Successor Landowner Suits for Recovery of Hazardous Waste Cleanup Costs: CERCLA Section 107 (a)(4)." *UCLA Law Review* 33 (August 1986): 1737–1775.

Deutsch, Stuart L., and others. "An Analysis of Regulations under the Resource Conservation and Recovery Act [dealing with the generation, transportation, treatment, storage, and disposal of hazardous wastes]." *Washington University Journal of Urban and Contemporary Law* 25 (1983): 145–202.

"Developments in the Law: Toxic Waste Litigation." *Harvard Law Review* 99 (May 1986): 1458–1661.

Dominguez, George S. "Current World Trends in Regulating Chemicals." *Toxic Substances Journal* 4 (Autumn 1982): 154–170.

Fisher, A., and others. "Communicating Risk under Title III of SARA: Strategies for Explaining Very Small Risks in a Community Context." *Journal of the Air Pollution Control Association* 39 (March 1989): 271–276.

Florio, James J. "Congress as Reluctant Regulator: Hazardous Waste Policy in the 1980s." *Yale Journal on Regulation* 3 (Spring 1986): 351–382.

Ginsburg, Richard A. "TSCA'S Unfulfilled Mandate for Comprehensive Regulation of Toxic Substances: The Potential of TSCA Section 21 Citizens' Petitions." *Environmental Law Reporter* 16 (November 1986): 10330–10337.

Goldberg, Paul. "Muzzling the Watchdog: EPA in Disarray." *Washington Monthly* 13 (December 1981): 30–35.

Gough, Michael. "Laws for the Regulation of Carcinogens: Identifying and Estimating the Risks That the Laws Seek To Reduce." *Toxic Substances Journal* 4 (Spring 1983): 251–276.

Gough, Robert G. "Workers' Right-To-Know about Chemical Hazards in the Workplace: A Proposed Model Uniform Right-To-Know Act and a Critical Look at Cincinnati's Right-To-Know Ordinance." *Northern Kentucky Law Review* 10:3 (1983): 427–460.

Guidotti, Tee L. "San Diego County's Community Right-To-Know Ordinance: Case Study of a Local Approach to Hazardous Substances Control." *Journal of Public Health Policy* 5 (September 1984): 396–409.

Harrington, Arthur J. "The Right to a Decent Burial: Hazardous Waste and Its Regulation in Wisconsin." *Marquette Law Review* 66 (Winter 1983): 223–279.

Haynes, Gerald D. "The Constitutionality of Trade Secret Disclosure Pursuant to the Toxic Substances Control Act of 1976." *IDEA* 27:2 (1986): 135–147.

Koplin, Allen N. "Right-To-Know: Implications of New Jersey's Law." *Journal of Public Health Policy* 5 (December 1984): 538–549.

Kriz, Margaret E. "Fuming over Fumes: Under a Community Right-To-Know Law, the Local Officials Are Starting To Learn about Stockpiles and Emissions of Dangerous Chemicals, But What Will They Do with the Data?" *National Journal* 20 (November 26, 1988): 3006–3009.

Latin, Howard. "Good Science, Bad Regulation, and Toxic Risk Assessment." *Yale Journal on Regulation* 5 (Winter 1988): 89–148.

Lennett, David J., and Linda E. Greer. "State Regulation of Hazardous Waste." *Ecology Law Quarterly* 12:2 (1985): 183–269.

McClellan, James W. "Hazardous Substances and the Right To Know in Canada." *International Labour Review* 128:5 (1989): 639–650.

McSlarrow, Kyle E. "The Department of Defense Environmental Cleanup Program: Application of State Standards to Federal Facilities after SARA." *Environmental Law Reporter* 17 (April 1987): 10120–10127.

Martens M., and others. "Some Thoughts on a Possible Regulatory Approach at EEC Level on the Classification and Labelling of Dangerous Preparations." *Toxic Substances Journal* 6 (Summer 1984): 44–60.

Mendeloff, John. "Does Overregulation Cause Underregulation? The Case of Toxic Substances." *Regulation [American Enterprise Institute]* 5 (September–October 1981): 47–52.

Montague, Peter. "What Must We Do: A Grass-Roots Offensive against Toxics in the 90s." *Workbook [Southwest Research and Information Center]* 14 (July–September 1989): 90–113.

New Jersey. Department of Environmental Protection. Division of Environmental Quality. *New Jersey Worker and Community Right To Know Act: Right To Know Hazardous Substance List.* Trenton, NJ: Right To Know Program, Division of Occupational and Environmental Health, New Jersey Department of Health, 1988. 134p.

New Jersey. Department of Health. *Report to the Governor and the Legislature of the State of New Jersey on the Implementation of the Worker and Community Right To Know Act, 1983–1987.* Trenton, NJ: 1987. 106p.

New Jersey. Senate. Energy and Environment Committee. *Public Hearing on Community Right-To-Know and Chemical Safety Act: Held: April 30, 1985, Trenton, New Jersey.* Trenton, NJ: 1985. 130p.

O'Reilly, James T. "Driving a Soft Bargain: Unions, Toxic Materials, and Right To Know Legislation." *Harvard Environmental Law Review* 9:2 (1985): 307–329.

Olschewsky, D., and A. Megna. "Hazardous-Waste Regulations Summarized for Refiners." *Oil and Gas Journal* 86 (January 4, 1988): 39–44.

Piasecki, B., and J. Gravander. "The Missing Links: Restructuring Hazardous-Waste Controls in America." *Technology Review* 88 (October 1985): 42–52.

Rebovich, Donald J. "Policing Hazardous Crime: The Importance of Regulatory/Law Enforcement Strategies and Cooperation in Offender Identification and Prosecution." *Criminal Justice Quarterly* 9 (Fall 1987): 173–215.

———. *Understanding Hazardous Waste Crime: A Multistate Examination of Offense and Offender Characteristics in the Northeast.* Trenton, NJ: Department of Law and Public Safety, Division of Criminal Justice, 1986. 70p.

Rosbe, William L., and Robert L. Gulley. "The Hazardous and Solid Waste Amendments of 1984: A Dramatic Overhaul of the Way America Manages Its Hazardous Wastes." *Environmental Law Reporter* 14 (December 1984): 10458–10467.

Rosenblatt, Jean. "Compensating Victims of Toxic Substances." *Editorial Research Reports* (October 15, 1982): 759–772.

Savage, Betty. "New Labelling Requirements within the EEC." *Toxic Substances Journal* 5 (Autumn 1983): 109–137.

Shortreed, J. H., and A. Stewart. "Risk Assessment and Legislation." *Journal of Hazardous Materials* 20 (December 1988): 315–334.

Smith, Janet D. "Private Monitoring of Hazardous Waste Sites: A Primer on Subsection 3013 Orders [under the United States Resource Conservation and Recovery Act]." *Environmental Law Reporter* 14 (May 1984): 10202–10208.

Stanfield, Rochelle L. "Few Are Satisfied with Statutes Aimed at Controlling the Chemical Revolution: Environmentalists, Consumer Activists and Some Members of Congress Blame Bureaucratic Footdragging, But the Bureaucrats Say Their Task Is an Impossible One." *National Journal* 16 (November 17, 1984): 2200–2205.

Staples, Charles A., and A. Frances Werner. "Priority Pollutant Assessment in the U.S.A.: Scientific and Regulatory Implications." *Toxic Substances Journal* 6 (Spring–Summer 1985): 186–200.

Strand, Palma J. "The Inapplicability of Traditional Tort Analysis to Environmental Risks: The Example of Toxic Waste Pollution Victim Compensation." *Stanford Law Review* 35 (February 1983): 575–619.

Susser, Peter A. "Chemical Hazard Disclosure Obligations: All Manufacturing Employers Are Now Obligated To Identify Hazardous Chemicals in Their Work Places and To Provide Employees with Information about Them." *Employment Relations Today* 13 (Winter 1986/1987): 301–308.

Tapscott, Gail. "Community Right-To-Know: A New Environmentalist Agenda." *Environmental Forum* 2 (October 1984): 8–14.

U.S. Congress. House. Committee on Energy and Commerce. Subcommittee on Oversight and Investigations. *Hazardous Waste Enforcement: Report, December 1982.* 97th Cong., 2d sess. Washington, DC: GPO, 1982. 43p.

U.S. Congress. House. Committee on Science, Space, and Technology. Subcommittee on International Science Cooperation. *The International Competitive*

Implications of Toxicological Standards: The Need for Consistent International Standards: Hearing, March 17, 1988. 100th Cong., 2d sess. Washington, DC: GPO, 1988. 229p.

U.S. Congress. Senate. Committee on Environment and Public Works. *A Legislative History of the Comprehensive Environmental Response, Compensation and Liability Act of 1980 (Superfund), Public Law 96-510, Together with a Section by Section Index.* 97th Cong., 2d sess. Washington, DC: GPO, 1983. 3 vols.

U.S. Laws, Statutes. *The Comprehensive Environmental Response, Compensation, and Liability Act of 1980 (Superfund) (P.L. 96-510), December 1986: As Amended by the Superfund Amendments and Reauthorization Act of 1986 (P.L. 99-499).* 99th Cong., 2d sess. Washington, DC: GPO, 1987. 226p.

Wasserman, Ursula. "Attempts at Control over Toxic Waste [legislation in Europe, the United States, and Canada, and international efforts]." *Journal of World Trade Law* 15 (September–October 1981): 410–430.

Wellenreuther, G. "Japanese Chemical Laws." *Toxic Substances Journal* 8:1 (1988): 45–62.

"What Are the Practical Implications of Proposition 65?" *Environmental Forum* 5 (May–June 1988): 16–20.

Williams, Bruce A., and Albert R. Matheny. "Testing Theories of Social Regulation: Hazardous Waste Regulation in the American States." *Journal of Politics* 46 (May 1984): 428–458.

Wurth-Hough, Sandra J. "Chemical Contamination and Governmental Policy Making: The North Carolina Experience." *State and Local Government Review* 14 (May 1982): 54–60.

"Your Right To Know: A New Law Is Revolutionizing How Companies, Communities and Governments Deal with Dangerous Chemicals." *Environmental Action* 20 (September–October 1988): 21–28.

LIABILITY

Agthe, Donald E. "Indemnity for Companies Adversely Affected by Environmental Regulation Changes." *Policy Studies Review* 6 (August 1986): 9–13.

Baker, R. Lisle, and Michael J. Markoff. "By-Product Liability: Using Common Law Private Actions To Clean Up Hazardous Waste Sites." *Harvard Environmental Law Review* 10:1 (1986): 99–134.

Bartlett, Kenneth G. "The Legal Development of a Viable Remedy for Toxic Pollution Victims." *Toxic Substances Journal* 4 (Spring 1983): 277–289.

Bloom, Gordon F. "The Hidden Liability of Hazardous-Waste Cleanup: New Regulations on the Disposal and Cleanup of Hazardous Waste Promise To Have Far-Reaching Effects on American Business." *Technology Review* 89 (February–March 1986): 58–64 +.

Burcat, Joel R. "Environmental Liability of Creditors: Open Season on Banks, Creditors and Other Deep Pockets." *Banking Law Journal* 103 (November–December 1986): 509–541.

"Controlling Exposure to Potential Liability for Off-Site Disposal of Hazardous Wastes: A Report by the Committee on Environmental Controls." *Business Lawyer* 39 (November 1983): 307–354.

Dare, Michael. "The Standard of Civil Liability for Hazardous Waste Disposal Activity: Some Quirks of Superfund." *Notre Dame Lawyer* 57 (December 1981): 260–284.

DeMoss, Douglas P. "The Bankruptcy Code and Hazardous Waste Cleanup: An Examination of the Policy Conflict." *William and Mary Law Review* 27 (Fall 1985): 165–216.

Dorge, Carol L. "After 'Voluntary' Liability: The EPA's Implementation of Superfund [to finance the cleanup of hazardous waste sites]." *Boston College Environmental Affairs Law Review* 11 (April 1984): 443–478.

Drabkin, Murray, and others. "Bankruptcy and the Cleanup of Hazardous Waste: Caveat Creditor." *Environmental Law Reporter* 15 (June 1985): 10168–10184.

Dubuc, Carroll E., and William D. Evans, Jr. "Recent Developments under CERCLA: Toward a More Equitable Distribution of Liability." *Environmental Law Reporter* 17 (June 1987): 10197–10204.

Dworkin, Judith M. "Private Parties Rights To Recover Losses from Groundwater Contamination in Arizona." *Arizona State Law Journal* 1985:3 (1985): 727–761.

"The Fairness and Constitutionality of Statutes of Limitations for Toxic Tort Sites." *Harvard Law Review* 96 (May 1983): 1683–1702.

Green, Michael D. "The Paradox of Statutes of Limitations in Toxic Substances Litigation." *California Law Review* 76 (October 1988): 965–1014.

"Groundwater Liability Waivers." *Environmental Forum* 4 (January 1986): 28–35.

Hall, Ridgway M., Jr. "The Problems of Unending Liability for Hazardous Waste Management." *Business Lawyer* 38 (February 1983): 593–621.

Hinds, Richard deC. "Liability under Federal Law for Hazardous Waste Injuries." *Harvard Environmental Law Review* 6:1 (1982): 1–33.

James, Walter D., III. "Financial Institutions and Hazardous Waste Litigation: Limiting the Exposure to Superfund Liability." *Natural Resources Journal* 28 (Spring 1988): 329–355.

"Joint and Several Liability for Hazardous Waste Release under Superfund [liability allocation problem under the Comprehensive Environmental Response, Compensation and Liability Act of 1980]." *Virginia Law Review* 68 (May 1982): 1157–1195.

Kelly, Peter B. "Changes in the Ownership of Hazardous Waste Disposal Sites: Original and Successor Liability." *Marquette Law Review* 67 (Summer 1984): 691–729.

Kimble, James L. "Conflicting Trends in Toxic Tort Liability." *Environmental Claims Journal* 2 (Winter 1989/1990): 153–164.

"Liability Insurance against Environmental Damage: A Status Report, June 1982 [emphasis on hazardous waste pollution]." *Business Lawyer* 38 (November 1982): 217–239.

"Liability of Parent Corporation for Hazardous Waste Cleanup and Damages." *Harvard Law Review* 99 (March 1986): 986–1003.

Martin, Thomas J. "Long-Term Liability for Hazardous Waste-Induced Injury in Missouri: Latent Harm Sufferers Beware." *Washington University Journal of Urban and Contemporary Law* 28 (1985): 299–343.

Pawlow, Jonathan R. "Liability for Shipments by Sea of Hazardous and Noxious Substances." *Law and Policy in International Business* 17:2 (1985): 455–481.

Priesing, Charles P. "Environmental Impairment Liability Insurance—Is It an Endangered Species?" *Toxic Substances Journal* 6 (Autumn–Winter 1984): 127–140.

Rich, David A. "Personal Liability for Hazardous Waste Cleanup: An Examination of CERCLA Section 107." *Boston College Environmental Affairs Law Review* 13 (1986): 643–671.

Shaw, Dianna Baker. "United States–Based Multinational Corporations Should Be Tried in the United States for Their Extraterritorial Toxic Torts." *Vanderbilt Journal of Transnational Law* 19 (Summer 1986): 651–670.

Shea, Edward E. "Protecting the Lender against Environmental Risk." *Toxic Substances Journal* 8:1 (1988): 33–43.

Smith, Turner T., Jr. "Environmental Damage Liability Insurance: A Primer." *Business Lawyer* 39 (November 1983): 333–354.

Spector, Morgan. "Compensating Victims of Hazardous Wastes." *Glendale Law Review* 6 (1984): 125–160.

Steinbeck, Margaret O. "Liability of Defense Contractors for Hazardous Waste Cleanup Costs." *Military Law Review* 125 (Summer 1989): 55–97.

Stever, Donald W. "Perspectives on the Problems of Federal Facility Liability for Environmental Contamination." *Environmental Law Reporter* 17 (April 1987): 10114–10119.

Trauberman, Jeffrey. "Statutory Reform of 'Toxic Torts': Relieving Legal, Scientific and Economic Burdens on the Chemical Victim." *Harvard Environmental Law Review* 7:2 (1983): 177–296.

Ulen, Thomas S., and others. "Minnesota's Environmental Response and Liability Act: An Economic Justification." *Environmental Law Reporter* 15 (April 1985): 10109–10115.

Van Lieshout, John M. "Breaking the Bank: Liability under Superfund; Lending Institutions May Be Liable for the Environmental Clean-Up Costs of Mortgaged Properties." *Real Estate Review* 16 (Fall 1986): 51–55.

Wallace, Perry E., Jr. "Liability of Corporations and Corporate Officers, Directors, and Shareholders under Superfund: Should Corporate and Agency Law Concepts Apply?" *Journal of Corporation Law* 14 (Summer 1989): 839–888.

"Waste Not, Want Not, But Not in This Country: Will There Ever Be Enough Insurance To Cover Every Possible Pollution Liability Loss Exposure?" *Journal of American Insurance* 63 (Third Quarter 1987): 1–5.

Zeller, Susan T., and Lisa M. Burke. "Theories of State Recovery under CERCLA for Injuries to the Environment [Comprehensive Environmental Response, Compensation and Liability Act of 1980]." *Natural Resources Journal* 24 (October 1984): 1101–1115.

SUPERFUND

Acton, Jan Paul. *Understanding Superfund: A Progress Report.* Santa Monica, CA: Rand Corporation, 1989. 65p.

Atkeson, Timothy B., and others. "An Annotated Legislative History of the Superfund Amendments and Reauthorization Act of 1986 (SARA)." *Environmental Law Reporter* 16 (December 1986): 10363–10419.

Bass, Steven B. "The Impact of the 1986 Superfund Amendments and Reauthorization Act on the Commercial Lending Industry: A Critical Assessment." *University of Miami Law Review* 41 (March 1987): 879–910.

Blaymore, Amy. "Retroactive Application of Superfund: Can Old Dogs Be Taught New Tricks?" *Boston College Environmental Affairs Law Review* 12 (Fall 1985): 1–50.

Bleicher, Samuel A., and Benjamin G. Stonelake, Jr. "Caveat Emptor: The Impact of Superfund and Related Laws on Real Estate Transactions." *Environmental Law Reporter* 14 (January 1984): 10017–10024.

Brown, Donald A. "Superfund Cleanups, Ethics, and Environmental Risk Assessment." *Boston College Environmental Affairs Law Review* 16 (Winter 1988): 181–198.

Brown, Theodore G., III. "Superfund and the National Contingency Plan: How Dirty Is 'Dirty'? How Clean Is 'Clean'?" *Ecology Law Quarterly* 12:1 (1984): 89–147.

Carlson, J. Lon, and Charles W. Bausell, Jr. "Financing Superfund: An Evaluation of Alternative Tax Mechanisms." *Natural Resources Journal* 27 (Winter 1987): 103–122.

Davis, Joseph A. "Special Report: Hazardous Wastes: Superfund Contaminated by Partisan Politics [cleanup program of EPA]." *Congressional Quarterly Weekly Report* 42 (March 17, 1984): 615–620.

"The Environmental Superfund Controversy: Pro & Con." *Congressional Digest* 65 (June–July 1986): 163–192.

Freeman, George C., Jr. "Inappropriate and Unconstitutional Retroactive Application of Superfund Liability." *Business Lawyer* 42 (November 1986): 215–248.

Giltenan, Edward F. "Superfund: EPA Has Dispersed $1.6 Billion To Clean Up the Nation's Toxic Waste Sites—But the Agency Is Accused of Mismanaging Superfund Program and Accomplishing Little." *Environmental Action* 18 (September–October 1986): 12–15.

Glass, Elizabeth Ann. "Superfund and SARA: Are There Any Defenses Left?" *Harvard Environmental Law Review* 12:2 (1988): 385–463.

Habicht, F. H., II. "The Role of the Agency for Toxic Substances and Disease Registry under the Superfund Amendments and Reauthorization Act of 1986." *Journal of Hazardous Materials* 18 (June 1988): 219–227.

Hill, R. D., and R. A. Olexsey. "Overview of the Superfund Innovative Technology Evaluation (SITE) Program." *Journal of the Air Pollution Control Association* 39 (January 1989): 16–21.

Hirschhorn, Joel S., and Kirsten U. Oldenburg. "Are We Cleaning Up? An Assessment of Superfund." *Chemical Engineering Progress* 84 (December 1988): 55–65.

Jones, David E., and Kyle E. McSlarrow. "But Were Afraid To Ask: Superfund Case Law, 1981–1989." *Environmental Law Reporter* 19 (October 1989): 10430–10457.

Josephson, J. "Implementing Superfund [cleaning up abandoned hazardous waste sites]." *Environmental Science and Technology* 20 (January 1986): 23–28.

Leifer, Steven L., and others. "Caught in a Squeeze Play: Debunking Some Myths about the Superfund Enforcement Program." *Environmental Forum* 5 (November–December 1988): 6–10.

Longest, H. "Building Public Confidence in Superfund." *Journal of Water Pollution Control Federation* 61 (March 1989): 298–303.

Lucero, Gene A. "Son of Superfund: Can the Program Meet Expectations?" *Environmental Forum* 5 (March–April 1988): 5–12.

McNiel, Douglas W., and Andrew W. Foshee. "Superfund Financing Alternatives." *Policy Studies Review* 7 (Summer 1988): 751–760.

Mays, Richard H. "EPA's Superfund Settlement Policy." *Environmental Forum* 3 (February 1985): 6–17.

———. "Settlements with SARA: A Comprehensive Review of Settlement Procedures under the Superfund Amendments and Reauthorization Act." *Environmental Law Reporter* 17 (April 1987): 10101–10113.

Milch, Thomas H. "Lender Liability under Superfund: The Increasing Risks of Exposure." *Issues in Bank Regulation* 12 (Summer 1988): 3–7.

Moorman, James W. "Superfund Settlement Policy: Will It Work?" *Environmental Forum* 4 (June 1985): 7–16.

———. "The Superfund Steering Committee: A Primer." *Environmental Forum* 4 (February 1986): 13–20.

Murphy, Margaret. "The Impact of 'Superfund' and Other Environmental Statutes on Commercial Lending and Investment Activities." *Business Lawyer* 41 (August 1986): 1133–1163.

"The New Superfund: Protecting People and Their Environment." *EPA Journal* 13 (January–February 1987): 2–35.

Novick, Sheldon M. "What Is Wrong with Superfund? Is the Remedial Program So Geared to the Catastrophic Sites, the Love Canals, That It Cannot Accommodate Most of the Nation's 'Normal' Soil and Groundwater Contamination Problems? Ambient Standards and Minimum Treatment Levels May Be Needed over the Long Run." *Environmental Forum* 2 (November 1983): 6–11.

"Recovery for Exposure to Hazardous Substances: The Superfund [Subsection] 301(e) Study and Beyond." *Environmental Law Reporter* 14 (March 1984): 10098–10141.

"Superfund's Petroleum Exclusion." *Environmental Forum* 3 (August 1984): 32–38.

U.S. Congress. House. Committee on Energy and Commerce. Subcommittee on Commerce, Transportation, and Tourism. *Implementation of the Superfund Program: Hearings, November 10, 1983–January 25, 1984.* 98th Cong., 1st and 2d sess. Washington, DC: GPO, 1984. 612p.

———. *Superfund Oversight: Hearings, February 19–July 29, 1981, on Oversight of the Comprehensive Environmental Response, Compensation and Liability Act of 1980.* 97th Cong., 1st sess. Washington, DC: GPO, 1982. 267p.

U.S. Congress. House. Committee on Energy and Commerce. Subcommittee on Oversight and Investigations. *EPA (Environmental Protection Agency) Withholding of Superfund Files: Hearings, December 3 and 14, 1982, on Contempt of Congress Proceedings against Environmental Protection Agency Administrator Anne M. Gorsuch.* 97th Cong., 2d sess. Washington, DC: GPO, 1983. 298p.

———. *EPA Investigation of Superfund and Agency Abuses: Hearings, Pts. 1–3, February 17–September 28, 1983.* 98th Cong., 1st sess. Washington, DC: GPO, 1984. 3 pts.

———. *Investigation of the Environmental Protection Agency: Report, August 1984, on the President's Claim of Executive Privilege over EPA Documents, Abuse in the Superfund Program, and Other Matters: Together with Minority Views.* 98th Cong., 2d sess. Washington, DC: GPO, 1984. 318p.

U.S. Congress. Joint Committee on Taxation. *Background and Issues Relating to the Reauthorization and Financing of the Superfund: Scheduled for Hearings before the Committee on Finance on April 25 and 26, 1985.* Washington, DC: GPO, 1985. 69p.

U.S. Congress. Office of Technology Assessment. *Are We Cleaning Up? 10 Superfund Case Studies: A Special Report of OTA's Assessment on Superfund Implementation.* Washington, DC: GPO, 1988. 76p.

————. *Coming Clean: Superfund Problems Can Be Solved.* Washington, DC: GPO, 1989. 223p.

U.S. Congress. Senate. Committee on Environment and Public Works. *Superfund Oversight: Hearing, Pts. 1–2, February 23–April 8, 1983.* 98th Cong., 1st sess. Washington, DC: GPO, 1983. 2 pts.

U.S. Congress. Senate. Committee on Environment and Public Works. Subcommittee on Superfund and Environmental Oversight. *Superfund Implementation Hearings, April 14–July 23, 1987.* 100th Cong., 1st sess. Washington, DC: GPO, 1987. 609p.

U.S. Congress. Senate. Committee on Environment and Public Works. Subcommittee on Superfund, Ocean, and Water Pollution. *Oversight of the Environmental Protection Agency's Management Review of the Superfund Program: Hearing, June 15, 1989.* 101st Cong., 1st sess. Washington, DC: GPO, 1989. 160p.

U.S. Congress. Senate. Committee on Finance. *Superfund Issues: Hearings, September 19 and 21, 1984.* 98th Cong., 2d sess. Washington, DC: GPO, 1985. 763p.

Vollmann, Alan P. "Double Jeopardy: Lender Liability under Superfund." *Real Estate Law Journal* 16 (Summer 1987): 3–19.

Werner, James D. "The Cost of Superfund: How Much?" *Environmental Forum* 3 (October 1984): 15–21.

Yang, Edward, and James Tracy. "Superfund Technology: A Long-Run Perspective." *Environmental Forum* 4 (January 1986): 40–46.

Management

Basta, N., and others. "Chemical Enginers Speak Out on Hazardous-Waste Management." *Chemical Engineering* 92 (September 16, 1985): 58–69.

Bowman, Ann O'M. "Hazardous Waste Management: An Emerging Policy Area within an Emerging Federalism." *Publius* 15 (Winter 1985): 131–144.

Bowman, Ann O'M., and James P. Lester. "Hazardous Waste Management: State Government Activity or Passivity?" *State and Local Government Review* 17 (Winter 1985): 155–161.

Burke, Ann E. "A New, Integrated Approach to the Management of Industrial Wastes." *Industrial Wastes* 29:5 (September–October 1983): 18–19.

Caldwell, L. K. "Perpetual Care: The Role of Engineers in Management of Toxic Waste." *Journal of Professional Issues in Engineering* 112 (April 1986): 107–117.

Choi, Yearn H., and Peter S. Daley. "Hazardous Waste Management Initiatives in DOD [U.S. Department of Defense]." *Defense Management Journal* 19 (Fourth Quarter 1983): 30–37.

"Controlling Hazardous Waste." *EPA Journal* 10 (October 1984): entire issue.

Coppock, Rob. "Control of Chemical Hazards: Risk Analysis, Conceptual Outlooks, and Political Traditions in Selected Countries." *Journal of Public and International Affairs* 5 (Winter 1984): 67–81.

Cromwell, J. E. "Hazardous Waste Site Management." *Consulting Engineer* 66 (April 1986): 35–40.

Freeman, H. M. "Hazardous Waste Management: A Strategy for Environmental Improvement." *Journal of the Air Pollution Control Association* 38 (January 1988): 59–62.

"Hazardous Waste Management." *Consulting Engineer* 66 (May 1986): 37–48.

"Hazardous Waste Management." *Public Management* 68 (March 1986): 3–14.

Hirschhorn, J. S. "Cutting Production of Hazardous Waste." *Technology Review* 91 (April 1988): 52–61.

"Improved Hazardous Wastes Management Needs." *Chemical Engineering Progress* 82 (September 1986): 29–34.

Jennings, A. A., and P. Suresh. "Risk Penalty Functions for Hazardous Waste Management." *Journal of Environmental Engineering* 112 (February 1986): 105–122.

Katzman, Martin T. "Environmental Risk Management through Insurance." *Cato Journal* 6 (Winter 1987): 775–799.

Mackie, J. A., and K. Niesen. "Hazardous Waste Management: The Alternatives." *Chemical Engineering* 91 (August 6, 1984): 50–64.

Mazmanian, Daniel, and David Morell. "The Elusive Pursuit of Toxics Management." *Public Interest (Washington)* (Winter 1988): 81–98.

"Monsanto's Early Warning System: How a Major Chemical Company Manages the Risks and Benefits in Technological Progress." *Harvard Business Review* 59 (November–December 1981): 107–122.

Mumme, Stephen P. "Dependency and Interdependence in Hazardous Waste Management along the U.S.-Mexico Border." *Policy Studies Journal* 14 (September 1985): 160–168.

New Jersey. Department of Environmental Protection. Division of Hazardous Waste Management. *Hazardous Waste Mangement in New Jersey: A Guide to Rules, Programs and Officials*. Trenton, NJ: 1987. 43p.

O'Brien, Robert M., and others. "Open and Closed Systems of Decision Making: The Case of Toxic Waste Management." *Public Administration Review* 44 (July–August 1984): 334–340.

Truax, Hawley. "The Fortunes of Hazardous Waste: The Hazardous Waste Management Industry Is Booming—But What Are Those Profits Buying Society?" *Environmental Action* 19 (July–August 1987): 10–14.

U.S. Congress. House. Committee on Science and Technology. Subcommittee on Natural Resources, Agriculture Research, and Environment. *Hazardous Waste Management: Hearing, May 16, 1983*. 98th Cong., 1st sess. Washington, DC: GPO, 1984. 168p.

Vogel, G. A. "Air Emission Control at Hazardous Waste Management Facilities." *Journal of the Air Pollution Control Association* 35 (May 1985): 558–566.

Wolf, K. "Source Reduction and the Waste Management Hierarchy." *Journal of the Air Pollution Control Association* 38 (May 1988): 681–686.

Sites

Andreen, William L. "Defusing the 'Not in My Back Yard' Syndrome: An Approach to Federal Preemption of State and Local Impediments to the Siting of PCB Disposal Facilities." *North Carolina Law Review* 63 (June 1985): 811–847.

Bacow, Lawrence S., and James R. Milkey. "Overcoming Local Opposition to Hazardous Waste Facilities: The Massachussetts Approach." *Harvard Environmental Law Review* 6:2 (1982): 265–305.

Brant, William M. "Dade County's Site Assessment and Cleanup Program." *Site Selection Handbook* 32 (August 1987): 829–831.

Carnes, R. A. "A Hazardous Waste Siting Experience." *Journal of Hazardous Materials* 13 (April 1986): 227–232.

Crawford, Mark. "Hazardous Waste: Where To Put It?" *Science* 235 (January 9, 1987): 156–157.

Davis, Joseph A. "Nuclear Waste: An Issue That Won't Stay Buried." *Congressional Quarterly Weekly Report* 45 (March 14, 1987): 451–456.

Duffy, Celeste P. "State Hazardous Waste Facility Siting: Easing the Process through Local Cooperation and Preemption." *Boston College Environmental Affairs Law Review* 11 (October 1984): 755–804.

Farago, Klra, and others. "Not in My Town: Conflicting Views on the Siting of a Hazardous Waste Incinerator." *Risk Analysis* 9 (December 1989): 463–471.

Hancock, Don. "The Wasting of America: Target—Nevada, Target—New Mexico." *Workbook [Southwest Research and Information Center]* 13 (January–March 1988): 2–12.

Harris, Jeffrey S. "Toxic Waste Uproar: A Community History [interactions between public health officials and citizens involving a chemical waste site in Memphis, TN]." *Journal of Public Health Policy* 4:2 (1983): 181–201.

Holznagel, Bernd. "Negotiation and Mediation: The Newest Approach to Hazardous Waste Facility Siting." *Boston College Environmental Affairs Law Review* 13 (Spring 1986): 329–378.

Landers, Robert K. "Living with Hazardous Wastes: The Evacuation of Hundreds of Families from Love Canal in August 1978 Dramatized the Hazardous Waste Threat: Congress Responded by Creating the 'Superfund,' but Relatively Few Hazardous Waste Sites Have Actually Been Cleaned Up Despite Efforts to Strengthen the Program." *Editorial Research Reports* (July 29, 1988): 378–387.

LaVo, Carl. "Not in My Backyard." *National Wildlife* 26 (April–May 1988): 24–27.

Lazarus, Arthur G. "Guidelines for Reusing Closed Industrial Wastes Disposal Sites." *Industrial Wastes* 29:5 (September–October 1983): 12–14.

McQuaid-Cook, J., and K. J. Simpson. "Siting a Fully Integrated Waste

Management Facility." *Journal of the Air Pollution Control Association* 36 (September 1986): 1031–1036.

Morell, David. "The Siting of Hazardous Waste Facilities in California." *Public Affairs Report* 25 (October 1984): 1–10.

New Jersey. General Assembly. Environmental Quality Committee. *Public Meeting: Status of Civilian and Military Cleanup of Hazardous Discharge Sites: Trenton, New Jersey, June 25, 1987.* Trenton, NJ: 1987. 148p.

————. *Public Meeting: Status of Civilian and Military Cleanup of Hazardous Discharge Sites: Trenton, New Jersey, May 21, 1987.* Trenton, NJ: 1987. 231p.

Olivieri, A. W., and others. "Groundwater Contamination in Site Rank Methodology." *Journal of Environmental Engineering* 112 (August 1986): 757–769.

Payne, B. A., and others. "The Effects on Property Values of Proximity to a Site Contaminated with Radioactive Waste." *Natural Resources Journal* 27 (Summer 1987): 579–590.

Portney, Kent E. "The Potential Theory of Compensation for Mitigating Public Opposition to Hazardous Waste Treatment Facility Siting: Some Evidence from Five Massachusetts Communities." *Policy Studies Journal* 14 (September 1985): 81–89.

"Siting of Hazardous Waste Facilities." *American Economic Review* 76 (May 1986): 285–299.

"Siting of Hazardous Waste Facilities and Transport of Hazardous Substances." *Environmental Law Reporter* 15 (August 1985): 10233–10261.

United Church of Christ. Commission for Racial Justice. *Toxic Wastes and Race in the United States: A National Report on the Racial and Socio-Economic Characteristics of Communities with Hazardous Waste Sites.* New York: UCC, 1987. 69p.

U.S. Congress. House. Committee on Energy and Commerce. Subcommittee on Commerce, Transportation, and Tourism. *Guidelines for Hazardous Waste Treatment Sites: Hearing, August 5, 1986.* 99th Cong., 2d sess. Washington, DC: GPO, 1986. 144p.

U.S. Congress. House. Committee on Interior and Insular Affairs. Subcommittee on Energy and the Environment. *Implementation of the Nuclear Waste Policy Act (Site Selection Program): Oversight Hearing, July 31, 1986.* 99th Cong., 2d sess. Washington, DC: GPO, 1987. 190p.

U.S. National Institute for Occupational Safety and Health. Divison of

Standards Development and Technology Transfer. *Hazardous Waste Sites and Hazardous Substance Emergencies*. Washington, DC: GPO, 1982. 22p.

Visocki, Kathryn. "Siting Controversial Facilities." *Popular Government* 55 (Summer 1989): 34–38.

Wells, Donald T. "Site Control of Hazardous Waste Facilities [Georgia]." *Policy Studies Review* 1 (May 1982): 728–735.

Zirschky, J. H., and D. J. Harris. "Geostatistical Analysis of Hazardous Waste Site Data." *Journal of Environmental Engineering* 112 (August 1986): 770–784.

Toxic and Hazardous Waste Disposal

General

Benson, R. J., Jr., and others. "Hazardous Waste Disposal as Concrete Admixture." *Journal of Environmental Engineering* 111 (August 1985): 441–447.

Cheyney, A. C. "Experience with the Co-Disposal of Hazardous Waste with Domestic Waste." *Chemistry and Industry* no. 17 (September 3, 1984): 609–615.

Claridge, F. B., and others. "Mine Waste Dumps Constructed in Mountain Valleys." *CIM Bulletin* 79 (August 1986): 79–87.

Hahn, Robert W. "An Evaluation of Options for Reducing Hazardous Waste." *Harvard Environmental Law Review* 12:1 (1988): 201–230.

Hillman, T. C. "Hazardous Waste Disposal and the Re-Use of Contaminated Land: The Waste Disposal Authority Role." *Chemistry and Industry* no. 17 (September 3, 1984): 602–609.

Hollod, G. J., and R. F. McCartney. "Waste Reduction in the Chemical Industry: Du Pont's Approach." *Journal of the Air Pollution Control Association* 38 (February 1988): 174–179.

Opplet, E. T. "Hazardous Waste Destruction [thermal techniques]." *Environmental Science and Technology* 20 (April 1986): 312–318.

Palmark, M. "Future Options for Disposal of Hazardous Wastes." *Chemistry and Industry* no. 12 (June 16, 1986): 416–422.

Piasecki, Bruce, and Gary A. Davis. "A Grand Tour of Europe's Hazardous Waste Facilities: Using a Combination of New Technology and Sophisticated Political Management, European Nations Have Avoided Many of the Pitfalls

Encountered by the United States in Dealing with Toxic Wastes." *Technology Review* 87 (July 1984): 20–29.

Rozich, A. F., and A. F. Gaudy, Jr. "Critical Point Analysis for Toxic Waste Treatment." *Journal of Environmental Engineering* 110 (June 1984): 562–572; Discussion, 111 (June 1985): 394–399.

Sather, N. F. "Hazardous Waste: Where To Put It? Where Will It Go?" *Mechanical Engineering* 110 (September 1988): 70–75.

Senkan, Selim M., and Nancy W. Stauffer. "What To Do with Hazardous Waste." *Technology Review* 84 (November–December 1981): 34–47.

Torrance, G. W., and D. Krewski. "Economic Evaluation of Toxic Chemical Control Programs." *Toxic Substances Journal* 7:1–4 (1985/1986): 53–71.

U.S. Congress. House. Committee on Armed Services. Investigations Subcommittee. *Army Disposal of Chemical Weapons: Hearing, July 25, 1986.* 99th Cong., 2d sess. Washington, DC: GPO, 1986. 314p.

U.S. Congress. House. Committee on Government Operations. Environment, Energy, and Natural Resources Subcommittee. *Potential Hazardous Waste Volume and Capacity Problems: Hearing, September 24, 1986.* 99th Cong., 2d sess. Washington, DC: GPO, 1987. 147p.

U.S. Congress. House. Committee on Science and Technology. Subcommittee on Investigations and Oversight. *Hazardous Waste Disposal: Hearings, March 30 and May 4, 1983.* 98th Cong., 1st sess. Washington, DC: GPO, 1983. 412p.

U.S. Congress. Office of Technology Assessment. *Serious Reduction of Hazardous Waste: For Pollution Prevention and Industrial Efficiency.* Washington, DC: GPO, 1986. 254p.

U.S. Environmental Protection Agency. Office of Solid Waste and Emergency Response. *The Hazardous Waste System.* Washington, DC: GPO, 1987. 5p.

————. *National Survey of Hazardous Waste Generators and Treatment, Storage and Disposal Facilities Regulated under RCRA [Resource Conservation and Recovery Act] in 1981.* Washington, DC: GPO, 1984. 232p.

Cleanup

Altmann, J. C. "A Regional Response for the 'Valley' [Phoenix, AZ, metropolitan area]." *Fire Command* 55 (October 1988): 30–35.

Gravitz, N. "Derivation and Implementation of Air Criteria during Hazardous Waste Site Cleanups." *Journal of the Air Pollution Control Association* 35 (July 1985): 753–758.

Hileman, B. "The Great Lakes Cleanup Effort." *Chemical and Engineering News* 66 (February 8, 1988): 22–39.

New Jersey. General Assembly. Agriculture and Environment Committee. *Public Hearing on Hazardous Waste Cleanup Operations: Held: Trenton, New Jersey, March 23, 1983.* Trenton, NJ: 1983. 291p.

New Jersey. Hazardous Waste Advisory Council. *Analysis of New Jersey's Hazardous Waste Site Cleanup Program.* Trenton, NJ: 1985. 5p.

Parker, A., and others. "The Decontamination of a Disused Gas Works Site." *Journal of Hazardous Materials* 9 (December 1984): 347–354.

Patchin, Peter J. "Valuation of Contaminated Properties." *Appraisal Journal* 56 (January 1988): 7–16.

Peaff, George, Jr. "Toxic Terrain: New Jersey May Be One of the Smallest States, but It's Also among the Most Contaminated: Is Enough Being Done To Clean Up This Toxic Mess?" *Business Journal of New Jersey* 7 (February 1990): 36–42 +.

Redding, Martin J. "Remedial Action Management Oversight—An Approach for Controlling Hazardous-Waste Cleanup Costs." *Toxic Substances Journal* 7:1–4 (1985/1986): 97–108.

Redding, Martin J., and George E. Howell. "Decision Framework for Cleanup Resolution." *Toxic Substances Journal* 5 (Spring 1984): 261–274.

"Strict Rules Dampen Water Cleanup." *Engineering News-Record* 217 (October 2, 1986): 36–44.

Trost, Cathy. "Hooker Chemical's Michigan Mess: Another Tale of Poisoned Wells and Dying Lakes, and Regulators Who Looked the Other Way." *Business and Society Review* (Winter 1981/1982): 32–39.

Disposal Technology

GENERAL

Allen, C. C., and B. L. Blaney. "Techniques for Treating Hazardous Wastes To Remove Volatile Organic Constituents." *Journal of the Air Pollution Control Association* 35 (August 1985): 841–848.

Anastos, G. J., and others. "Innovative Technologies for Hazardous Waste Treatment." *Nuclear and Chemical Waste Management* 8:4 (1988): 269–281.

Blaney, B. L. "Treatment Technologies for Hazardous Wastes: Managing Solvent Wastes." *Journal of the Air Pollution Control Association* 36 (March 1986): 275–285.

Cheremisinoff, P. N. "New Strategies for Haz Waste Treatment and Disposal." *Pollution Engineering* 20 (April 1988): 64–71.

"Emerging Technologies for Treating Hazardous Wastes." *Journal of Hazardous Materials* 12 (November 1985): 127–205.

Freeman, H. M., and R. A. Olexsey. "Treatment Technologies for Hazardous Wastes: Treatment Alternatives for Dioxin Wastes." *Journal of the Air Pollution Control Association* 36 (January 1986): 67–75.

Ghassemi, M. "Innovative In Situ Treatment Technologies for Cleanup of Contaminated Sites [detoxifier]." *Journal of Hazardous Materials* 17 (February 1988): 189–206.

Ghassemi, M., and M. Haro. "Hazardous Waste Surface Impoundment Technology." *Journal of Environmental Engineering* 111 (October 1985): 602–617.

Grosse, D. W. "Treatment Technologies for Hazardous Wastes: Metal Bearing Hazardous Waste Streams." *Journal of the Air Pollution Control Association* 36 (May 1986): 603–614.

Lee, K. W., and others. "The Advanced Electric Reactor: A New Technology for Hazardous Waste Destruction." *Journal of Hazardous Materials* 12 (November 1985): 143–160.

MacNeil, J. C. "Membrane Separation Technologies for Treatment of Hazardous Wastes." *Critical Reviews in Environmental Control* 18:2 (1988): 91–131.

Olexsey, R. A. "Treatment Technologies for Hazardous Wastes." *Journal of the Air Pollution Control Association* 36 (January 1986): 66–75; 36 (March 1986): 275-285; 36 (April 1986): 403-409; 36 (May 1986): 603-614.

Tucker, S. P., and G. A. Carson. "Deactivization of Hazardous Chemical Wastes." *Environmental Science and Technology* 19 (March 1985): 215–220.

U.S. Congress. House. Committee on Science and Technology. Subcommittee on Investigations and Oversight. *Hazardous Waste Treatment Technology: Hearing, May 2, 1985.* 99th Cong., 1st sess. Washington, DC: GPO, 1985. 220p.

AERATION

Matsuda, A., and others. "Behavior of Nitrogen and Phosphorus during Batch Aerobic Digestion of Waste Activated Sludge—Continuous Aeration and Intermittent Aeration by Control of DO." *Water Research* 22 (December 1988): 1495–1501.

Polprasert, C., and H. S. Raghunandana. "Wastewater Treatment in a Deep Aeration Tank." *Water Research* 19:2 (1985): 257–264.

Ros, M., and others. "An Improved Method Using Respirography for the Design of Activated Sludge Aeration Systems." *Water Research* 22 (December 1988): 1483–1489.

Sackellares, R. W., and others. "Development of a Dynamic Aerated Lagoon Model." *Journal of Water Pollution Control Federation* 59 (October 1987): 877–883.

Stukenberg, J. R. "Physical Aspects of Surface Aeration Design." *Journal of Water Pollution Control Federation* 56 (September 1984): 1014–1021.

BIOLOGICAL TREATMENT

Abeliovich, A. "Biological Treatment of Chemical Industry Effluents by Stabilization Ponds." *Water Research* 19:12 (1985): 1497–1503.

Becker, C. D., and others. "Evaluating Hazardous Materials for Biological Treatment." *Nuclear and Chemical Waste Management* 5:3 (1985): 183–192.

Douglas, J. "Cleaning Up with Biotechnology." *EPRI Journal* 13 (September 1988): 14–21.

Klemetson, S. L., and M. E. Lang. "Treatment of Saline Wastewater Using a Rotating Biological Contractor." *Journal of Water Pollution Control Federation* 56 (December 1984): 1254–1259.

Schmutzler, D. W., and others. "Startup and Operation of a Full-Scale Anaerobic Treatment System at a Groundwood and Coated Paper Mill." *Tappi Journal* 71 (November 1988): 103–112.

Stuckey, D. C., and P. L. McCarty. "The Effect of Thermal Pretreatment on the Anaerobic Biodegradability and Toxicity of Waste Activated Sludge." *Water Research* 18:11 (1984): 1343–1353.

CHLORINATION

Shertzer, R. H. "The Water Quality Impacts of Chlorinating Papermill

Effluents." *Journal of Water Pollution Control Federation* 57 (February 1985): 172–180.

Watkins, C. H., and R. S. Hammerschlag. "The Toxicity of Chlorine to a Common Vascular Aquatic Plant." *Water Research* 18:8 (1984): 1037–1043.

COAGULATION

Knocke, W. R., and L. P. Jones. "Mixing Effects on Coal Processing Waste Treatment." *Journal of Environmental Engineering* 109 (December 1983): 1295–1319; Discussion, 111 (August 1985): 543–548.

McTernan, W. F., and others. "Removal of Toxic Materials from In Situ Tar Sand Process Water." *Journal of Energy Engineering* 113 (December 1987): 79–91.

Robinson, S. M., and others. "Low-Activity-Level Process Wastewaters: Treatment by Chemical Precipitation and Ion Exchange." *Journal of Water Pollution Control Federation* 60 (December 1988): 2120–2127.

Thiem, L. T., and E. A. Alkhatib. "In Situ Adaptation of Activated Sludge by Shock Loading To Enhance Treatment of High Ammonia Content Petrochemical Wastewater." *Journal of Water Pollution Control Federation* 60 (July 1988): 1245–1252.

FILTRATION

Aulenbach, D. B., and N. Meisheng. "Studies on the Phosphorus Removal from Treated Wastewater by Sand." *Journal of Water Pollution Control Federation* 60 (December 1988): 2089–2094.

Harrison, J. R., and G. T. Daigger. "A Comparison of Trickling Filter Media." *Journal of Water Pollution Control Federation* 59 (July 1987): 679–685; Discussion, 59 (October 1987): 915–918.

Harrison, J. R., G. T. Daigger, and others. "A Survey of Combined Trickling Filter and Activated Sludge Processes." *Journal of Water Pollution Control Federation* 56 (October 1984): 1073–1079.

Logan, B. E., and others. "A Fundamental Model for Trickling Filter Process Design." *Journal of Water Pollution Control Federation* 59 (December 1987): 1029–1042.

FLOTATION PROCESS

Bennett, G. F. "The Removal of Oil from Wastewater by Air Flotation: A Review." *Critical Reviews in Environmental Control* 18:3 (1988): 189–253.

Fuchs, E. W., and others. "Removal of Water-Borne Paint Solids from Paint Spray Water." *Journal of Coatings Technology* 60 (December 1988): 89–93.

Lovett, D. A., and S. M. Travers. "Dissolved Air Flotation for Abattoir Wastewater." *Water Research* 20 (April 1986): 421–426.

Nolan, B. T., and others. "Multi-Stage Air Flotation of Tar Sand Wastewater." *Journal of Environmental Engineering* 112 (April 1986): 14–24.

INCINERATION

Brunner, C. R., and C. H. Brown. "Hospital Waste Disposal by Incineration." *Journal of the Air Pollution Control Association* 38 (October 1988): 1297–1309.

Ditz, Daryl. "Interpretation of Need in U.S. Ocean Incineration Policy." *Marine Policy* 13 (January 1989): 43–55.

———. "The Risk of Hazardous Waste Spills from Incineration at Sea." *Journal of Hazardous Materials* 17 (February 1988): 149–167.

Fort, W. C., III, and L. W. R. Dicks. "Chlorinated Waste Incinerator Heat Recovery Boiler Corrosion." *Materials Performance* 25 (March 1986): 9–14.

Hershkowitz, Allen. "Burning Trash: How It Could Work." *Technology Review* 90 (July 1987): 26–34.

Ho, M. D., and M. G. Ding. "Field Testing and Computer Modeling of an Oxygen Combustion System at the EPA Mobile Incinerator." *Journal of the Air Pollution Control Association* 36 (September 1988): 1185–1191.

Kamlet, K. S. "The Case for Ocean Incineration of Hazardous Wastes." *Marine Technology Society Journal* 20 (March 1986): 42–47.

Penner, S. S., and others. "Waste Incineration and Energy Recovery." *Energy* 13 (December 1988): 845–851.

Ross, Dennis R. "Incineration of Containerized Hazardous Wastes." *Industrial Wastes* 29:6 (November–December 1983): 13–14.

Sweet, W. E., and others. "Hazardous Waste Incineration: A Progress Report." *Journal of the Air Pollution Control Association* 35 (February 1985): 138–143.

Tessitore, J., and F. L. Cross. "Incineration of Hospital Infectious Waste." *Pollution Engineering* 20 (November 1988): 83–88.

Trenholm, A., and others. "Total Mass Emissions from a Hazardous Waste Incinerator." *Journal of Hazardous Materials* 18 (April 1988): 99–106.

U.S. Congress. House. Committee on Energy and Commerce. Subcommittee on Commerce, Transportation, and Tourism. *Ocean-Based Incineration of Hazardous Wastes: Hearing, February 11, 1981, on Options Open to Disposing of Wastes That Have Been Accumulated by the United States.* 97th Cong., 1st sess. Washington, DC: GPO, 1981. 100p.

U.S. Congress. House. Committee on Government Operations. Environment, Energy, and Natural Resources Subcommittee. *Incineration of Hazardous Wastes at Sea: Hearing, July 12, 1984.* 98th Cong., 2d sess. Washington, DC: GPO, 1985. 343p.

U.S. Congress. House. Committee on Merchant Marine and Fisheries. *Incineration of Hazardous Waste at Sea: Hearing, December 7, 1983, before the Subcommittee on Fisheries and Wildlife Conservation and the Environment and the Subcommittee on Oceanography, on an Oversight Regarding the Incineration of Hazardous Waste at Sea as Part of the Overall Efforts of Congress To Confront the Problem of Managing Hazardous Waste.* 98th Cong., 1st sess. Washington, DC: GPO, 1984. 461p.

U.S. Congress. House. Committee on Merchant Marine and Fisheries. Subcommittee on Oceanography. *Incineration of Hazardous Waste at Sea: Hearing, November 17, 1987, on H.R.737.* 100th Cong., 1st sess. Washington, DC: GPO, 1988. 261p.

U.S. Congress. Senate. Committee on Environment and Public Works. Subcommittee on Environmental Pollution. *Ocean Incineration: Hearings, June 19 and July 17, 1985.* 99th Cong., 1st sess. Washington, DC: GPO, 1985. 593p.

U.S. Environmental Protection Agency. Office of Solid Waste and Emergency Response. *Final Environmental Impact Statement for the Offshore Platform Hazardous Waste Incineration Facility.* Washington, DC: GPO, 1982. 4 vols.

Velde, G. Vander, and E. F. Glod. "Ocean-Based Incineration of Chemical Waste: Background, Technology and Performance of the Vulcanus Systems." *Marine Technology Society Journal* 20 (March 1986): 29–37.

Walker, Christopher A. "The United States Environmental Protection Agency's Proposal for At-Sea Incineration of Hazardous Wastes: A Transnational Perspective." *Vanderbilt Journal of Transnational Law* 21:1 (1988): 157–189.

LAGOONS

Amy, G. L., and others. "Bisorption of Organic Halide in a Kraft Mill Generated Lagoon." *Journal of Water Pollution Control Federation* 60 (August 1988): 1445–1453.

Eischen, G. W., and J. D. Keenan. "Monitoring Aerated Lagoon Performance." *Journal of Water Pollution Control Federation* 57 (August 1985): 876–881.

Koerner, R. M., and A. E. Lord, Jr. "Spill Alert Device for Earth Dam Failure Warning." *Journal of Hazardous Materials* 9 (December 1984): 373–380.

Pearson, H. W., and others. "Guidelines for the Minimum Evaluation of the Performance of Full-Scale Waste Stabilization Pond Systems." *Water Research* 21 (September 1987): 1067–1075.

Robinson, H. D., and G. Grantham. "The Treatment of Landfill Leachates in On-Site Aerated Lagoon Plants: Experience in Britain and Ireland." *Water Research* 22 (June 1988): 733–747.

LIME TREATMENT

Aggour, M. S., and others. "Field Aging of Fixed Sulfur Dioxide Scrubber Waste." *Journal of Energy Engineering* 111 (September 1985): 62–73.

Day, D. R., and others. "Evaluation of Hazardous Waste Incineration in a Lime Kiln: Rockwell Lime Company." *Journal of Hazardous Materials* 12 (December 1985): 313–321.

MINIMIZATION

Drabkin, M. "The Waste Minimization Assessment: A Useful Tool for the Reduction of Industrial Hazardous Wastes." *Journal of the Air Pollution Control Association* 38 (December 1988): 1530–1541.

"Hazardous Waste Minimization: Hazardous Waste Minimization within the Department of Defense." *Journal of the Air Pollution Control Association* 38 (August 1988): 1042–1050; 38 (September 1988): 1745–1785.

Hunt, G. E. "Hazardous Waste Minimization: Waste Reduction in the Metal Finishing Industry." *Journal of the Air Pollution Control Association* 38 (May 1988): 672–680.

Lorton, G. A. "Hazardous Waste Minimization: Waste Minimization in the Paint and Allied Products Industry." *Journal of the Air Pollution Control Association* 38 (April 1988): 422–427.

Oman, D. E. "Hazardous Waste Minimization: Waste Minimization in the Foundry Industry." *Journal of the Air Pollution Control Association* 38 (July 1988): 932–940.

Williams, Alan C. "A Study of Hazardous Waste Minimization in Europe:

Public and Private Strategies To Reduce Production of Hazardous Waste." *Boston College Environmental Affairs Law Review* 14 (Winter 1987): 165–255.

WASTEWATER TREATMENT

Cheremisinoff, P. N. "Systems for Hazardous Wastewater Mangement." *Pollution Engineering* 20 (September 1988): 80–90.

Christensen, D. R., and J. A. Girick. "Understanding Primary Clarifiers for Plant Wastewater Systems." *Plant Engineering* 40 (June 26, 1986): 39–42.

Corder, G. D., and P. L. Lee. "Feedforward Control of a Wastewater Plant." *Water Research* 20 (March 1986): 301–309.

Desmukh, S. B. "Toxic Organic Wastewater Treatment through Solid Phase Transformation with Re-Use Potential." *Journal of Hazardous Materials* 9 (August 1984): 171–179.

Eckenfelder, W. W. "Wastewater Treatment." *Chemical Engineering* 92 (September 2, 1985): 60–74.

Gupta, S. K. "Nitrogenous Wastewater Treatment by Activated Algae." *Journal of Environmental Engineering* 111 (February 1985): 61–77.

Hannah, S. A., and others. "Comparative Removal of Toxic Pollutants: Six Wastewater Treatment Processes." *Journal of Water Pollution Control Federation* 58 (January 1986): 27–34.

Lankford, P. W., and others. "Reducing Wastewater Toxicity." *Chemical Engineering* 95 (November 7, 1988): 72–82.

Lehr, V. A. "When Domestic Sewage Becomes Industrial Waste." *Heating/Piping/Air Conditioning* 59 (December 1987): 93–97.

New Jersey. Senate. Special Committee To Study Coastal and Ocean Pollution. *Public Hearing: The Pretreatment of Industrial Waste Waters Prior to Discharge into Publicly Owned Treatment Works: Trenton, New Jersey, September 15, 1987.* Trenton, NJ: 1987. 138p.

Nichols, A. B. "L.A.'s Wastewater Treatment Plant: Victim or Villain?" *Journal of Water Pollution Control Federation* 59 (November 1987): 932–938; Discussion, 60 (February 1988): 150.

Oron, G., and others. "Wastewater Treatment and Renovation by Different Duckweed Species." *Journal of Environmental Engineering* 112 (April 1986): 247–263.

Petrasek, A. C., Jr., and I. J. Kugelman. "Metals Removals and Partitioning in Conventional Wastewater Treatment Plants." *Journal of Water Pollution Control Federation* 55 (September 1983): 1183–1190; Discussion, 56 (December 1984): 1286–1287; 57 (March 1985): 263–264.

Ramirez, E. A., and O. F. D'Alessio. "Design and Engineering of a Wastewater Pretreatment Facility." *Metal Finishing* 82 (November 1984): 15–20.

Reed, S. C. "Nitrogen Removal in Wastewater Stabilization Ponds." *Journal of Water Pollution Control Federation* 57 (January 1985): 39–45.

Rose, J. B. "Microbial Aspects of Wastewater Reuse for Irrigation." *Critical Reviews in Environmental Control* 16:3 (1986): 231–256.

Waltrip, G. D., and E. G. Snyder. "Elimination of Odor at Six Wastewater Treatment Plants." *Journal of Water Pollution Control Federation* 57 (October 1985): 1027–1032.

Weller, L. W., and others. "Wastewater Treatment in the Kansas City Metropolitan Area." *Journal of Water Pollution Control Federation* 57 (September 1985): 902–911.

Yakovlev, S. V., and E. V. Dvinskych. "Wastewater Sludge Treatment and Utilization in the U.S.S.R." *Journal of Testing and Evaluation* 14 (May 1986): 168–171.

ZIMMERMAN PROCESS

Cheremisinoff, P. N. "Thermal Treatment Technologies for Hazardous Wastes." *Pollution Engineering* 20 (August 1988): 50–55.

Heimbuch, J. A., and A. R. Wilhelmi. "Wet Air Oxidation—A Treatment Means for Aqueous Hazardous Waste Streams." *Journal of Hazardous Materials* 12 (November 1985): 187–200.

Kalam, A., and J. B. Joshi. "Regeneration of Spent Earth by Wet Oxidation." *Journal of the American Oil Chemists' Society* 65 (September 1988): 1536–1540.

Land and Ocean Disposal

GENERAL

Atkins, Kenneth M. "Land Application of Industrial Wastes: A Viable Alternative for Disposal [disbursing relatively thin layers of waste over large areas for environmentally acceptable assimilation]." *Economic Development Review* 2 (Winter 1984): 35–50.

Bakalian, Allan. "Regulation and Control of United States Ocean Dumping: A Decade of Progress, an Appraisal for the Future." *Harvard Environmental Law Review* 8:1 (1984): 193–256.

Bewers, J. M., and C. J. R. Garrett. "Analysis of the Issues Related to Sea Dumping of Radioactive Wastes." *Marine Policy* 11 (April 1987): 105–124.

Boehmer-Christiansen, Sonja. "An End to Radioactive Waste Disposal 'at Sea'?" *Marine Policy* 10 (April 1986): 119–131.

Brown, Michael. "The Lower Depths Underground Injection of Hazardous Wastes." *Amicus Journal* 7 (Winter 1986): 14–23.

Byrne, C. D., and others. "Measurements of the Dispersion of Liquid Industrial Waste Discharged into the Wake of a Dumping Vessel." *Water Research* 22 (December 1988): 1577–1584.

Eichholz, G. G., and others. "Migration of Elemental Mercury through Soil from Simulated Burial Sites." *Water Research* 22 (January 1988): 15–20.

Guarascio, John A. "The Regulation of Ocean Dumping after *City of New York v. Environmental Protection Agency.*" *Boston College Environmental Affairs Law Review* 12 (Summer 1985): 701–741.

Higgins, T. E., and others. "Avoiding Land Disposal of Hazardous Waste." *Journal of the Air Pollution Control Association* 37 (November 1987): 1331–1336.

Hunsaker, Carolyn T. "Ocean Dumping of Low-Level Radioactive Waste: Review of U.S. Laws and International Agreements." *Environmental Forum* 3 (November 1984): 24–31.

Lahey, William L. "Economic Changes for Environmental Protection: Ocean Dumping Fees [United States]." *Ecology Law Quarterly* 11:3 (1984): 305–342.

————. "Ocean Dumping of Sewage Sludge: The Tide Turns from Protection to Management [regulation efforts by EPA]." *Harvard Environmental Law Review* 6:2 (1982): 395–431.

Lavelle, J. W. "Settling Speeds of Sewage Sludge in Seawater." *Environmental Science and Technology* 22 (October 1988): 1201–1207.

Lechich, A. F., and F. J. Roethel. "Marine Disposal of Stabilized Metal Processing Waste." *Journal of Water Pollution Control Federation* 60 (January 1988): 93–99.

Miller, Leonard A., and Robert S. Taylor. "The Enemy Below: EPA Plans Action on Leaking Underground Storage Tanks." *Environmental Law Reporter* 15 (May 1985): 10135–10143.

New Jersey. Senate. Special Committee To Study Coastal and Ocean Pollution. *Public Hearing: Testimony Concerning the Various Sources of Ocean Pollution, Including Sludge and Dredge Spoil Dumping, Vessel Refuse, and Other Ocean Dumping Practices: Long Branch, New Jersey, January 7, 1987.* Trenton, NJ: 1987. 325p.

Spiller, Judith, and Cynthia Hayden. "Radwaste at Sea: A New Era of Polarization or a New Basis for Consensus?" *Ocean Development and International Law* 19:5 (1988): 345–366.

Spirer, Julian H. "The Ocean Dumping Deadline: Easing the Mandate Millstone [burden of federal rules and regulations]." *Fordham Urban Law Journal* 11:1 (1982/1983): 1–49.

Stoddard, A., and others. "Development and Application of a Deepwater Ocean Waste Disposal Model for Dredged Material: Yabucoa Harbor, Puerto Rico." *Marine Technology Society Journal* 19:3 (1985): 26–39.

Symons, B. D., and others. "Fate and Transport of Organics in Soil: Model Predictions and Experimental Results." *Journal of Water Pollution Control Federation* 60 (Summer 1988): 1684–1693.

U.S. Congress. House. Committee on Merchant Marine and Fisheries. *Ocean Dumping: Hearing, February 23, 1988, before the Subcommittee on Oceanography and the Subcommittee on Fisheries and Wildlife Conservation and the Environment.* 100th Cong., 2d sess. Washington, DC: GPO, 1988. 362p.

————. *Ocean Dumping: Hearing, Pt.2, March 1, 1984, before the Subcommittee on Oceanography and the Subcommittee on Fisheries and Wildlife Conservation and the Environment on H.R.4829 [and other bills].* 98th Cong., 2d sess. Washington, DC: GPO, 1984, 352p.

————. *Plastic Pollution in the Marine Environment: Hearings, June 17 and July 23, 1987, before the Subcommittee on Coast Guard and Navigation and the Subcommittee on Fisheries and Wildlife Conservation and the Environment on H.R.940, A Bill To Provide for the Regulation of the Disposal of Plastic Materials and Other Garbage at Sea; To Provide for Negotiation, Regulation, and Research Regarding Fishing with Plastic Driftnets.* 100th Cong., 1st sess. Washington, DC: GPO, 1987. 498p.

————. *Waste Dumpings: Hearings, May 1–November 5, 1981, before the Subcommittee on Oceanography and the Subcommittee on Fisheries and Wildlife Conservation and the Environment on Title I, Marine Protection, Research and Sanctuaries Act; Ocean*

Dumping and Dumping Deadline; Radioactive Waste Dumping; Land Based Alternatives to Ocean Dumping. 97th Cong., 1st sess. Washington, DC: GPO, 1982. 544p.

U.S. Congress. House. Committee on Merchant Marine and Fisheries. Subcommittee on Coast Guard and Navigation. *Plastic Pollution in the Marine Environment: Hearing, August 12, 1986, on the Problem of Nonbiodegradable Plastic Refuse in the Marine Environment, and To Examine the Options That Exist on All Levels for Responding to It.* 99th Cong., 2d sess. Washington, DC: GPO, 1986. 210p.

U.S. Congress. Office of Technology Assessment. *Wastes in Marine Environments.* Washington, DC: GPO, 1987. 313p.

Van Dyke, Jon M. "Ocean Disposal of Nuclear Wastes." *Marine Policy* 12 (April 1988): 82–95.

Zeppetello, Marc A. "National and International Regulation of Ocean Dumping: The Mandate To Terminate Marine Disposal of Contaminated Sewage Sludge." *Ecology Law Quarterly* 12:3 (1985): 619–664.

LANDFILLS

Ainsworth, J. B., and A. O. Ojeshina. "Specify Containment Liners." *Hydrocarbon Processing* 63 (November 1984): 130–135.

Albaiges, J., and others. "Organic Indicators of Groundwater Pollution by a Sanitary Landfill." *Water Research* 20 (September 1986): 1153–1159.

Baker, L. W., and K. P. MacKay. "Screening Models for Estimating Toxic Air Pollution near a Hazardous Waste Landfill." *Journal of the Air Pollution Control Association* 35 (November 1985): 1190–1195.

Bonistall, D. F., and D. E. Oman. "A Fast-Track Approach through the Design and Permit Phases for an Ash Landfill." *Tappi Journal* 71 (August 1988): 85–90.

Daniel, D. E., and C. D. Shackelford. "Disposal Barriers That Release Contaminants Only by Molecular Diffusion." *Nuclear and Chemical Waste Management* 8:4 (1988): 299–305.

Harrop-Williams, K. "Clay Liner Permeability: Evaluation and Variation." *Journal of Geotechnical Engineering* 111 (October 1985): 1211–1225; Discussion, 113 (August 1987): 943–949.

Hillman, T. C. "Hazardous Waste Disposal and the Re-Use of Contaminated Land: The Waste Disposal Authority Role." *Chemistry and Industry* no. 17 (September 3, 1984): 602–609.

Hillman, T. C., P. W. Jayawickrama, and others. "Leakage Rates through Flaws in Membrane Liners." *Journal of Environmental Engineering* 114 (December 1988): 1401–1420.

Hulick, Betty. *Landfill Gas Recovery Seminar, Rutgers University [New Brunswick, NJ], November 16, 1981: Methane Saves Money; Methane Makes Money.* Trenton, NJ: Department of Environmental Protection, Solid Waste Administration, 1981. 5p.

Johnson, R. R., and others. "Diffusive Contaminant Transport in Natural Clay: A Field Example and Implications for Clay-Land Waste Disposal Sites." *Environmental Science and Technology* 23 (March 1989): 340–349.

McEnroe, B. M., and P. R. Schroeder. "Leachate Collection in Landfills: Steady Case." *Journal of Hydraulic Engineering* 114 (October 1988): 1051–1062.

Mahmood, R. J. "Mobility of Organics in Land Treatment Systems." *Journal of Environmental Engineering* 112 (April 1986): 236–245.

Mosher, Lawrence. "Who's Afraid of Hazardous Dumps? Not Us, Says the Reagan Administration; The Environmental Protection Agency Thinks Landfills Can Be Made Secure, but Environmentalists and Even Parts of the Chemical Industry Don't Agree." *National Journal* 14 (May 29, 1982): 952–957.

Parra, J. O. "Electrical Response of a Leak in a Geomembrane Liner." *Geophysics* 53 (November 1988): 1445–1452.

Peirce, J. J. "Overburden Pressures Exerted on Clay Liners." *Journal of Environmental Engineering* 112 (April 1986): 280–291.

Peirce, J. J., and K. A. Witter. "Termination Criteria for Permeability Testing." *Journal of Geotechnical Engineering* 112 (September 1986): 841–854.

Peirce, J. J., and others. "Clay Liner Construction and Quality Control." *Journal of Environmental Engineering* 112 (February 1986): 13–24.

Peyton. R. L., and P. R. Schroeder. "Field Verification of HELP Model for Landfills." *Journal of Environmental Engineering* 114 (April 1988): 247–269.

Robinson, H. D., and P. J. Maris. "The Treatment of Leachates from Domestic Waste in Landfill Sites." *Journal of Water Pollution Control Federation* 57 (January 1985): 30–38.

Russell, G. M., and A. L. Higer. "Assessment of Ground-Water Contamination near Lantana Landfill, Southeast Florida." *Ground Water* 26 (March–April 1988): 156–164.

Spencer, L. LeSeur, and L. D. Drake. "Hydrogeology of an Alkaline Fly Ash Landfill in Eastern Iowa." *Ground Water* 25 (September–October 1987): 519–526.

Valsaraj, K. T., and L. J. Thibodeaux. "Equilibrium Adsorption of Chemical Vapors on Surface Soils, Landfills and Landforms—A Review." *Journal of Hazardous Materials* 19 (July 1988): 79–99.

Weber, W. J., Jr. "Organic Contamination: Whistling Past the Graveyard." *Journal of Water Pollution Control Federation* 58 (January 1986): 12–17.

Wood, J. A., and M. L. Porter. "Hazardous Pollutants in Class II Landfills." *Journal of the Air Pollution Control Association* 37 (May 1987): 609–615.

LOVE CANAL

Albanese, Jay S. "Love Canal Six Years Later: The Legal Legacy in Determining Responsibility for Health Problems Experienced in the Niagara Falls, New York, Area and Relocation and Cleanup Costs." *Federal Probation* 48 (June 1985): 533–538.

Brown, Michael H. "Love Canal Revisited: Ten Years Later There Are Still More Questions than Answers." *Amicus Journal* 10 (Summer 1988): 37–44.

Brummer, James. "Love Canal and the Ethics of Environmental Health." *Business and Professional Ethics Journal* 2 (Summer 1983): 1–25.

Danzo, Andrew. "The Big Sleazy: Love Canal Ten Years Later." *Washington Monthly* 20 (September 1988): 11–14 +.

Deegan, J., Jr. "Looking Back at Love Canal." *Environmental Science and Technology* 21 (April 1987): 328–331; (May 1987): 421–426.

U.S. Congress. House. Committee on Energy and Commerce. Subcommittee on Commerce, Transportation, and Tourism. *Love Canal Study and Habitability Statement: Hearing, August 9, 1982.* 97th Cong., 2d sess. Washington, DC: GPO, 1983. 226p.

Worthley, John A., and Richard Torkelson. "Managing the Toxic Waste Problem: Lessons from the Love Canal [Niagara Falls, NY]." *Administration and Society* 13 (August 1981): 145–160.

Transportation

Belardo, S. "Information Support for Control of Hazardous Materials Movement." *Journal of Hazardous Materials* 10 (February 1985): 13–32.

Cain, Jonathan T. "Routes and Roadblocks: State Controls on Hazardous Waste Imports." *Natural Resources Journal* 23 (October 1983): 767–793.

Cantilli, Edmund J., and Raymond D. Scanlon. "Death on Wheels; Hazardous Materials Transport: An Explosive National Problem." *Journal of Insurance* 42 (July–August 1981): 2–7.

Carnes, S. A. "Institutional Issues Affecting the Transport of Hazardous Materials in the United States: Anticipating Strategic Management Needs." *Journal of Hazardous Materials* 13 (August 1986): 257–277.

Coleman, J. A. "Railroad-Highway Crossings and Route Selection for Transporting Hazardous Materials." *Public Roads* 48 (September 1984): 63–71.

De Bivre, Aline F. M. "Liability and Compensation for Damage in Connection with the Carriage of Hazardous and Noxious Substances by Sea." *Journal of Maritime Law and Commerce* 17 (January 1986): 61–88.

Eddy, Ronald M., and Diana Terry Riendl. "Transporter Liability under CERCLA." *Environmental Law Reporter* 16 (September 1986): 10244–10254.

Gold, Edgar. "Legal Aspects of Transportation of Dangerous Goods at Sea." *Marine Policy* 10 (July 1986): 185–191.

Haddow, George D. "The Safe Transportation of Hazardous Materials." *Transportation Quarterly* 41 (July 1987): 309–322.

Kalevela, S. A., and A. E. Radwan. "International Issues of Transporting Hazardous Materials." *Transportation Quarterly* 42 (January 1988): 125–139.

Krause, Kitry. "Toxic Chemicals and 18-Wheelers: The Dangers of Deregulation; Economic Deregulation without Tough Enforcement of Safety Rules Is a Recipe for Disaster." *Washington Monthly* 17 (November 1985): 39–44.

List, G., and M. Abkowitz. "Estimates of Current Hazardous Materials Flow Patterns." *Transportation Quarterly* 40 (October 1986): 483–502.

———. "Towards Improved Hazardous Materials Flow Data." *Journal of Hazardous Materials* 17 (March 1988): 287–304.

"Local Response to Hazardous Materials Incidents and Accidents." *Transportation Quarterly* 40 (October 1986): 461–482.

Marten, Bradley M. "Regulation of the Transportation of Hazardous Materials: A Critique and a Proposal." *Harvard Environmental Law Review* 5:2 (1981): 345–376.

New Jersey. Senate. Law, Public Safety and Defense Committee. *Public Hearing: To Elicit Testimony for Discussion on the Role of Local Governmental and Law Enforcement Officials on Monitoring the Transportation of Hazardous Materials: Paramus, New Jersey, May 1, 1988.* Trenton, NJ: 1988. 56p.

Peterson, Sybil, and others. "The Horrors of Hazardous Waste Hauling: Accidents Waiting To Happen." *Business and Society Review* (Winter 1987): 62–66.

Purdy, G., and others. "An Analysis of the Risks Arising from the Transport of Liquified Gases in Great Britain." *Journal of Hazardous Materials* 20 (December 1988): 335–355.

Rittvo, Steven M., and George D. Haddow. "Transportation of Hazardous Materials: A Case Study [St. Bernard Parish, LA]." *Transportation Quarterly* 38 (January 1984): 137–151.

Rothberg, Paul F. "Selected DOT Hazardous Materials Transportation Training Regulations and Options for Congressional Action." *Transportation Quarterly* 40 (October 1986): 451–460.

Rublack, Suzanne. "Controlling Transboundary Movements of Hazardous Waste: The Evolution of a Global Convention." *Fletcher Forum* 13 (Winter 1989): 113–125.

Starobin, Paul. "Safety Sought in Moving Hazardous Materials." *Congressional Quarterly Weekly Report* 45 (May 23, 1987): 1031–1036.

United Nations. Committee of Experts on the Transport of Dangerous Goods. *Recommendations on the Transport of Dangerous Goods: Test and Criteria.* New York: U.N. Agent, 1986. 189p.

————. *Recommendations on the Transport of Dangerous Goods: Test and Criteria.* 6th rev. ed. New York: U.N. Agent, 1989. 482p.

United Nations. Economic Commission for Europe. Inland Transport Committee. *European Agreement Concerning the International Carriage of Dangerous Goods by Road (ADR) and Protocol of Signature, Done at Geneva on 30 September 1957; Amendment 1.* New York: U.N. Agent, 1988. 116p.

U.S. Congress. House. *Mailing of Biological Toxins: Joint Hearings, June 23 and October 5, 1988, before the Subcommittee on Postal Personnel and Modernization of the Committee on Post Office and Civil Service and the Subcommittee on Research and Development of the Committee on Armed Services.* 100th Cong., 2d sess. Washington, DC: GPO, 1988. 192p.

U.S. Congress. House. Committee on Energy and Commerce. Subcommittee on Telecommunications, Consumer Protection, and Finance. *Motor Carrier Safety: Transportation of Hazardous and Nuclear Materials: Hearing, July 19, 1985.* 99th Cong., 1st sess. Washington, DC: GPO, 1986. 183p.

U.S. Congress. House. Committee on Interior and Insular Affairs. Subcommittee on Energy and the Environment. *Nuclear Waste Transportation: Hearing, May 12, 1988, on H.R.1649, To Establish a Requirement That No Person May Offer Any High-Level Radioactive Waste for Transportation in Interstate Commerce Unless Licensed for Such Offering by the Nuclear Regulatory Commission, and for Other Purposes.* 100th Cong., 2d sess. Washington, DC: GPO, 1988. 407p.

U.S. Congress. House. Committee on Public Works and Transportation. *Hazardous Materials Transportation: Joint Hearing, March 25, 1981, before the Subcommittee on Surface Transportation and the Subcommittee on Aviation.* 97th Cong., 1st sess. Washington, DC: GPO, 1982. 58p.

U.S. Congress. House. Committee on Public Works and Transportation. Subcommittee on Surface Transportation. *Transportation of Hazardous Materials: Hearings, May 19 and 25, 1988.* 100th Cong., 2d sess. Washington, DC: GPO, 1988. 847p.

U.S. Congress. Office of Technology Assessment. *Hazardous Materials Transportation: Hearing, April 22–May 12, 1987.* 100th Cong., 1st sess. Washington, DC: GPO, 1987. 375p.

———. *Transportation of Hazardous Materials.* Washington, DC: GPO, 1986. 265p.

———. *Transportation of Hazardous Materials: Summary.* Washington, DC: GPO, 1986. 55p.

U.S. Congress. Senate. Committee on Commerce, Science, and Transportation. Subcommittee on Surface Transportation. *Hazardous Materials Transportation Act, Natural Gas Pipeline Safety Act, and Hazardous Liquid Pipeline Safety Act Authorizations: Hearings, April 28, 1981, on S.960.* 97th Cong., 1st sess. Washington, DC: GPO, 1981. 99p.

U.S. Congress. Senate. Committee on Governmental Affairs. Subcommittee on Energy, Nuclear Proliferation, and Government Processes. *Enforcement of Federal Regulations and Penalties for Shipments of Hazardous and Radioactive Materials: Hearing, May 9, 1984.* 98th Cong., 2d sess. Washington, DC: GPO, 1984. 169p.

U.S. Department of Transportation. *Transportation of Hazardous Materials.* Washington, DC: GPO, 1980. 50p.

U.S. National Transportation Safety Board. Office of Evaluations and Safety Objectives. *Safety Effectiveness Evaluation: Federal and State Enforcement Efforts in Hazardous Materials Transportation by Truck*. Springfield, VA: National Technical Information Service, 1981. 109p.

Pesticides

General

Araki, F. "Chemical Control of Plant Diseases in Japan." *Chemistry and Industry* no. 2 (January 20, 1986): 54–60.

Baker, Brian P. "Pest Control in the Public Interest: Crop Protection in California." *UCLA [University of California, Los Angeles] Journal of Environmental Law and Policy* 8:1 (1989): 31–71.

Daniels, R. W. "Pesticides—A Classic Case of Poor PR?" *Chemistry and Industry* no. 22 (November 21, 1988): 712–717.

Dover, Michael. "Getting Off the Pesticide Treadmill." *Technology Review* 88 (November–December 1985): 52–61.

Fujita, Y. "Current and Future Problems of Insect Control in Japan." *Chemistry and Industry* no. 2 (January 20, 1986): 61–67.

Hansen, Michael. "Pesticides: No Way To Feed the Hungry." *WHY Magazine* (Summer 1989): 24–27 +.

Herrett, R. A. "Low-Dosage Pesticides: Less Is Better." *Chemtech* 18 (April 1988): 220–225.

Jones, P. "The Public Perception of Pesticide Risk." *ASTM Standardization News* 16 (March 1988): 48–53.

MacIntyre, Angus A. "Why Pesticides Received Extensive Use in America: A Political Economy of Agricultural Pest Management to 1970." *Natural Resources Journal* 27 (Summer 1987): 533–578.

McLean, J. E., and others. "Evaluation of Mobility of Pesticides in Soil Using U.S. EPA Methodology." *Journal of Environmental Engineering* 114 (June 1988): 689–703.

Mattes, Kitty. "Kicking the Pesticide Habit: Farmers Have a Long Way To Go, but Experts See Hope in Integrated Pest Management, Organic Farming, and Low-Impact Sustainable Agriculture." *Amicus Journal* 11 (Fall 1989): 10–17.

Raheel, M. "Resistance of Selected Textiles to Pesticide Penetration and Degradation." *Journal of Environmental Health* 42 (January–February 1987): 214–219.

Sacks, Carolyn, and others. "Consumer Pesticide Concerns: A 1965 and 1984 Comparison." *Journal of Consumer Affairs* 21 (Summer 1987): 96–107.

Sherman, J. "Pesticides [review of applications]." *Analytical Chemistry* 57:5 (April 1985): 1R–15R.

Stanfield, Rochelle L. "Politics Pushes Pesticide Manufacturers and Environmentalists Closer Together: Reacting to Political Blackmail, a Manufacturers Group Reached Agreement with an Environmental-Consumer Coalition after Blocking Pesticide Control Reform for Years." *National Journal* 17 (December 14, 1985): 2846–2851.

Strock, W. J. "Demand for Home and Garden Pesticides Spurs New Products." *Chemical and Engineering News* 65 (April 6, 1987): 11–16.

———. "Pesticides Growth Slows." *Chemical and Engineering News* 65 (November 16, 1987): 35–42.

Thomas, M. S. "The Pesticide Market in Brazil." *Chemistry and Industry* no. 6 (March 21, 1988): 179–184.

U.S. Congress. House. Committee on Energy and Commerce. Subcommittee on Oversight and Investigations. *Pesticides in Food: Hearing, April 30, 1987.* 100th Cong., 1st sess. Washington, DC: GPO, 1987. 317p.

U.S. Congress. House. Committee on Government Operations. Environment, Energy, and Natural Resources Subcommittee. *EPA's Pesticide Indemnification and Disposal Program: Hearing, July 9, 1987.* 100th Cong., 1st sess. Washington, DC: GPO, 1988. 153p.

U.S. Congress. House. Committee on Small Business. Subcommittee on Energy and Agriculture. *Generic Pesticide Registration: Hearing, June 24, 1987.* 100th Cong., 1st sess. Washington, DC: GPO, 1987. 139p.

———. *Pesticide Registration Fees: Hearings, November 3 and 5, 1987.* 100th Cong., 1st sess. Washington, DC: GPO, 1988. 169p.

"Use of Pesticides To Control Post-Harvest Losses of Food Crops." *Chemistry and Industry* no. 3 (February 4, 1985): 70–90.

Whitney, R. W., and others. "Flow of Vegetable Oil-Pesticide Blank-Formulation Mixtures through Agricultural Spray Nozzles." *Journal of the American Oil Chemists' Society* 63 (March 1986): 340–345.

Younos, T. M., and D. L. Weigmann. "Pesticides: A Continuing Dilemma." *Journal of Water Pollution Control Federation* 60 (July 1988): 1199–1205.

Dioxin

Belton, Thomas J., and others. *A Study of Dioxin (2,3,7,8-Tetrachlorodibenzo-p-dioxin) Contamination in Select Finfish, Crustaceans and Sediments of New Jersey Waterways.* Trenton, NJ: Department of Environmental Protection, Office of Science and Research, 1985. 102p.

Freestone, F., and others. "Evaluation of On-Site Incineration for Cleanup of Dioxin Contaminated Materials." *Nuclear and Chemical Waste Management* 7:1 (1987): 3–20.

Hanson, D. J. "Science Failing To Back Up Veteran Concerns about Agent Orange." *Chemical and Engineering News* 65 (November 9, 1987): 7–11 +; Discussion, 66 (January 18, 1988): 2.

Jacobs, James B., and Dennis McNamara. "Vietnam Veterans and the Agent Orange Controversy." *Armed Forces and Society* 13 (Fall 1986): 57–79.

Myers, G. L., and D. G. Patterson. "Safety and Operations Procedures for Handling Dioxins and Other Chemical Toxicants." *Professional Safety* 32 (June 1987): 30–37.

Shaub, W. M., and W. Tsang. "Dioxin Formation in Incinerators." *Environmental Science and Technology* 17 (December 1983): 721–730.

Tschirley, F. H. "Dioxin." *Scientific American* 254 (February 1986): 29–35; Discussion, 254 (April 1986): 4 +.

U.S. Congress. House. Committee on Veterans' Affairs. Subcommittee on Hospitals and Health Care. *Agent Orange Studies: Hearing, July 31, 1986.* 99th Cong., 2d sess. Washington, DC: GPO, 1986. 159p.

Weisenhahn, D. F., and others. "A Simplified Model for Dioxin and Furan Formation in Municipal-Waste Incinerators." *Energy* 13 (March 1988): 225–237.

Environmental Aspects

Borrelli, Peter. "Pesticide Alert." *Amicus Journal* 10 (Spring 1988): 20–29.

Bruchet, A., and others. "Continuous Composite Sampling and Analysis of Pesticides in Water." *Water Research* 18:11 (1984): 1401–1409.

Lau, L. S., and J. F. Mink. "Organic Contamination of Groundwater: A Learning Experience." *American Water Works Association Journal* 79 (August 1987): 37–43.

Redfield, Sarah E. "Chemical Trespass? An Overview of Statutory and Regulatory Efforts To Control Pesticide Drift." *Kentucky Law Journal* 73:3 (1984/1985): 855–918.

Rice, C. P., and others. "Atmospheric Transport of Taxapene to Lake Michigan." *Environmental Science and Technology* 20 (November 1986): 1109–1116.

Thiel, D. A., and others. "The Effects of a Sludge Containing Dioxin on Wildlife in Pine Plantations." *Tappi Journal* 72 (January 1989): 94–99.

U.S. Congress. Senate. Committee on Environment and Public Works. *Environmental Issues Related to the Use of Pesticides: Hearing, June 10, 1988.* 100th Cong., 2d sess. Washington, DC: GPO, 1988. 136p.

Vick, W. H., and others. "Physical Stabilization Technique for Mitigation of Environmental Pollution from Dioxin Contaminated Soils." *Journal of Hazardous Materials* 18 (May 1988): 189–206.

Export-Import Trade

Chetley, Andrew. "Not Good Enough for Us but Fit for Them: An Examination of the Chemical and Pharmaceutical Export Trades." *Journal of Consumer Policy* 9 (June 1986): 155–180.

Goldberg, Karen A. "Efforts To Prevent Misuse of Pesticides Exported to Developing Countries: Progressing beyond Regulation and Notification." *Ecology Law Quarterly* 12:4 (1985): 1025–1051.

Hill, Raymond. "Problems and Policy for Pesticide Exports to Less Developed Countries." *Natural Resources Journal* 28 (October 1988): 699–720.

"Report Urges Better Chemical Controls Abroad." *Conservation Foundation Letter* no. 2 (1988): 1–7.

U.S. Congress. House. Committee on Government Operations. Environment, Energy, and Natural Resources Subcommittee. *The Uncontrolled Export of Unregistered Pesticides: Hearing, May 3, 1989.* 101st Cong., 1st sess. Washington, DC: GPO, 1989. 267p.

Health Effects

Branson, D. H., and others. "Thermal Response Associated with Prototype Pesticide Protective Clothing." *Textile Research Journal* 56 (January 1986): 27–34.

Davis, Joseph A. "Concern over Tainted Food Could Propel Pesticides Bill: New Studies Raise Safety Questions." *Congressional Quarterly Weekly Report* 45 (June 13, 1987): 1269–1272.

Fenske, R. A., and others. "A Video Imaging Technique for Assessing Dermal Exposure." *American Industrial Hygiene Association Journal* 47 (December 1986): 764–775.

Meyerhoff, Albert H., and Lawrie Mott. "Another Man's Poison: Pesticides, Cancer, and You." *Amicus Journal* 7 (Fall 1985): 28–34.

"Pesticides and the Consumer." *EPA Journal* 13 (May 1987): 2–43.

Stellman, S. D., and others. "Health and Reproductive Outcome among American Legionnaires in Relation to Combat and Herbicide Exposure in Vietnam." *Environmental Research* 47 (December 1988): 150–174.

Stone, J. F., and others. "Relationships between Clothing and Pesticide Poisoning—Symptoms among Iowa Farmers." *Journal of Environmental Health* 50 (January–February 1988): 210–215.

U.S. Congress. House. Committee on Agriculture. Subcommittee on Department Operations, Research and Foreign Agriculture. *Pesticide Food Safety Act of 1988: Hearings, July 28 and September 7, 1988, on H.R.4937.* 100th Cong., 2d sess. Washington, DC: GPO, 1988. 192p.

U.S. Congress. House. Committee on Energy and Commerce. Subcommittee on Health and the Environment. *Dioxin Contamination of Food and Water: Hearing, December 8, 1988.* 100th Cong., 2d sess. Washington, DC: GPO, 1989. 216p.

U.S. Congress. House. Committee on Veterans' Affairs. Subcommittee on Hospitals and Health Care. *Scientific Research on the Health of Vietnam Veterans: Hearing, June 8, 1988.* 100th Cong., 2d sess. Washington, DC: GPO, 1988. 461p.

————. *Studies on Agent Orange: Hearing, July 10, 1989.* 101st Cong., 1st sess. Washington, DC: GPO, 1989. 284p.

Wright, Angus. "Rethinking the Circle of Poison: The Politics of Pesticide Poisoning among Mexican Farm Workers." *Latin American Perspectives* 13 (Fall 1986): 26–59.

Yang, R. S. H. "A Toxicological View of Pesticides." *Chemtech* 17 (November 1987): 698–703.

Herbicides

Abbott, I. M., and others. "Worker Exposure to a Herbicide Applied with Ground Sprayers in the United Kingdom." *American Industrial Hygiene Association Journal* 48 (February 1987): 167–175.

Benbrook, Charles M., and Phyllis B. Moses. "Engineering Crops To Resist Herbicides: We Can Reduce Our Dependence on Cancer-Causing Herbicides by Breeding Crops That Resist More Benign Compounds." *Technology Review* 89 (November–December 1986): 54–61.

Buzzo, V. "Herbicides in Brazil." *Chemistry and Industry* no. 6 (March 21, 1988): 190–195.

Fenton, R. "Herbicide Use in Major Arable Crops." *Chemistry and Industry* no. 6 (March 20, 1989): 167–172.

Libich, S., and others. "Occupational Exposure of Herbicide Applicators to Herbicides Used along Electric Power Transmission Line Right-of-Way." *American Industrial Hygiene Association Journal* 45 (January 1984): 56–62.

Snel, M. "Past-Emergence Herbicides for Broad-Leaved Weed Control in Cereals." *Chemistry and Industry* no. 6 (March 20, 1989): 172–177.

Stellman, S. D., and others. "Combat and Herbicide Exposures in Vietnam among a Sample of American Legionnaires." *Environmental Research* 47 (December 1988): 112–128.

Laws and Regulations

Bosso, Christopher J. "Transforming Adversaries into Collaborators: Interest Groups and the Regulation of Chemical Pesticides." *Policy Sciences [Amsterdam]* 21:1 (1988): 3–22.

Ferguson, Scott, and Ed Gray. "1988 FIFRA Amendments: A Major Step in Pesticide Regulation." *Environmental Law Reporter* 19 (February 1989): 10070–10082.

Nicholson, Chester A. "Agent Orange Products Liability Litigation." *Air Force Law Review* 24:2 (1984): 97–124.

Sand, Robert H. "How Much Is Enough? Observations in Light of the Agent Orange Settlement." *Harvard Environmental Law Review* 9:2 (1985): 283–306.

Thrupp, Lori Ann. "Pesticides and Policies: Approaches to Pest-Control Dilemmas in Nicaragua and Costa Rica." *Latin American Perspectives* 15 (Fall 1988): 37–70.

U.S. Congress. House. Committee on Energy and Commerce. *PCB and Dioxin Cases: Hearing, November 19, 1982.* 97th Cong., 2d sess. Washington, DC: GPO, 1983. 370p.

U.S. Congress. House. Committee on Energy and Commerce. Subcommittee on Health and the Environment. *Food Safety Amendments of 1989: Hearings, May 17 and 31, 1989, on H.R.1725, a Bill To Amend the Federal Food, Drug, and Cosmetic Act To Revise the Authority under That Act To Regulate Pesticide Chemical Residues in Food.* 101st Cong., 1st sess. Washington, DC: GPO, 1989. 773p.

————. *Regulation of Pesticide Residues: Hearing, July 17, 1986.* 99th Cong., 2d sess. Washington, DC: GPO, 1987. 263p.

————. *Regulation of Pesticides: Hearing, June 8, 1987.* 100th Cong., 1st sess. Washington, DC: GPO, 1988. 239p.

U.S. Congress. House. Committee on Energy and Commerce. Subcommittee on Oversight and Investigations. *EPA Contracting: Pesticide Regulation: Hearing, May 14, 1987.* 100th Cong., 1st sess. Washington, DC: GPO, 1987. 215p.

U.S. Congress. House. Committee on Environment and Public Works. Subcommittee on Toxic Substances, Environmental Oversight, Research, and Development. *Chemical and Food Crops: Hearing, May 15, 1989.* 101st Cong., 1st sess. Washington, DC: GPO, 1989. 256p.

————. *Government Regulation of Pesticides in Food: The Need for Administrative and Regulatory Reform: Report, October 1989.* 101st Cong., 1st sess. Washington, DC: GPO, 1989. 97p.

U.S. Congress. Senate. Committee on Environment and Public Works. Subcommittee on Hazardous Wastes and Toxic Substances. *The PCB Control Act of 1988: Hearing, August 11, 1988, on S.2693.* 100th Cong., 2d sess. Washington, DC: GPO, 1988. 141p.

Residues

Kawano, M., and others. "Bioconcentration and Residue Patterns of Chlordane Compounds in Marine Animals: Invertebrates, Fish, Mammals, and Seabirds." *Environmental Science and Technology* 22 (July 1988): 792–797.

Miltner, R. J., and others. "Treatment of Seasonal Pesticides in Surface Waters." *American Water Works Association Journal* 81 (January 1989): 43–52.

Petersen, B., and C. Chaisson. "Pesticides and Residues in Food." *Food Technology* 42 (July 1988): 59–64.

Stone, J. F., and others. "Laundering Pesticide-Soiled Clothing: A Survey of Iowa Farm Families." *Journal of Environmental Health* 48 (March–April 1986): 259–264.

U.S. Congress. Office of Technology Assessment. *Pesticide Residues in Food: Technologies for Detection.* Washington, DC: GPO, 1988. 232p.

Oil Spills

General

Chamberlain, G. "Technology Tackles the Oil Spill [MunMAP]." *Design News* 45 (June 19, 1989): 90–95.

Lacy, J. D. "How Ashland Oil Made the Best of an Unfortunate Situation." *American Mining Congress Journal* 74 (August 1988): 7–10.

Levine, R. A. "Contingency Planning for Oil Spills: An Update Based on Experience Gained from the Port Angeles Spill." *Marine Technology* 25 (April 1988): 145–159; Discussion, 26 (January 1989): 44–46.

New Jersey. Senate. Energy and Environment Committee. *Public Hearing: Oil Spill Prevention and Response Capability: Camden, New Jersey, April 19, 1989.* Trenton, NJ: 1989. 114p.

Nichols, A. B. "Alaskan Oil Spill Shocks the Nation." *Journal of Water Pollution Control Federation* 61 (July 1989): 1174–1185.

Nulty, Peter. "The Future of Big Oil: Is Exxon's Muck-up at Valdez a Reason To Bar Drilling in One of the Industry's Hottest Prospects? Not According to Those Closest to the Scene: The Alaskans." *Fortune* 119 (May 8, 1989): 46–49.

Organization for Economic Cooperation and Development. *Combatting Oil Spills: Some Economic Aspects*. Paris, France: OECD, 1982. 140p.

――――. *The Cost of Oil Spills*. Paris, France: OECD, 1982. 252p.

Prokop, J. "The Ashland Tank Collapse." *Hydrocarbon Processing* 67 (May 1988): 105–108.

Rubenstein, Daniel. "Black Oil, Red Ink: The Exxon Spill Proves It's Time To Move from Entrepreneurship to Ecopreneurship, but Accounting Still Has To Find a Way of Matching Revenues with Social Costs." *CA Magazine* 122 (November 1989): 28–34+.

Straube, Michele. "Is Full Compensation Possible for the Damages Resulting from the Exxon Valdez Oil Spill?" *Environmental Law Reporter* 19 (August 1989): 10338–10350.

U.S. Coast Guard. *Polluting Incidents in and around U.S. Waters, Calendar Year 1980 and 1981*. Washington, DC: Department of Transportation, Coast Guard, 1982. 41p.

U.S. Congress. House. Committee on Government Operations. Government Activities and Transportation Subcommittee. *Coast Guard Capabilities for Oil Spill Clean Up: Hearing, August 26, 1982*. 97th Cong., 2d sess. Washington, DC: GPO, 1983. 42p

U.S. Congress. House. Committee on Interior and Insular Affairs. Subcommittee on Water, Power, and Offshore Energy Resources. *Investigation of the Exxon Valdez Oil Spill, Prince William Sound, Alaska: Oversight Hearings: Pt. 1, May 5–8, 1989*. 101st Cong., 1st sess. Washington, DC: GPO, 1989. 1,145p.

U.S. Congress. House. Committee on Merchant Marine and Fisheries. Subcommittee on Coast Guard and Navigation. *Exxon Valdez Oil Spill: Hearing, April 6, 1989*. 101st Cong., 1st sess. Washington, DC: GPO, 1989. 291p.

――――. *Exxon Valdez Oil Spill Cleanup: Hearing, August 10, 1989, on Examining the Incident of the Exxon Valdez Oil Spill, the Cleanup Activities Performed, and To Learn Ways of Providing Federal Policy To Prevent Such Incidents in the Future*. 101st Cong., 1st sess. Washington, DC: GPO, 1989. 215p.

U.S. Congress. Senate. Committee on Commerce, Science, and Transportation. *Exxon Oil Spill: Hearing, Pt. 1, April 6, 1989, on Exxon Valdez Oil Spill and*

Its Environmental and Maritime Implications. 101st Cong., 1st sess. Washington, DC: GPO, 1989. 100p.

————. *Exxon Oil Spill: Hearings: Pt. 2, May 10 and July 20, 1989, before the National Ocean Policy Study and the Subcommittee on Merchant Marine, on Cleanup, Containment, and Impact of the Exxon Valdez Oil Spill and Oil Spill Prevention and Maritime Regulations.* 101st Cong., 1st sess. Washington, DC: GPO, 1989. 529p.

U.S. Congress. Senate. Committee on Environment and Public Works. Subcommittee on Environmental Protection. *Oil Spill on the Monongahela and Ohio Rivers: Hearing, February 4, 1988.* 100th Cong., 2d sess. Washington, DC: GPO, 1988. 139p.

————. *Oilspill in Prince William Sound, Alaska: Hearing, April 19, 1989.* 101st Cong., 1st sess. Washington, DC: GPO, 1989. 226p.

U.S. Congress. Senate. Committee on the Judiciary. Subcommittee on Administrative Practice and Procedure. *Oil Spill off Nantucket, Massachusetts: Hearing, December 22, 1976, on Damage Caused by the Oil Spill of the Ship the 'Argo Merchant,' December 15, 1976.* 94th Cong., 2d sess. Washington, DC: GPO, 1977. 252p.

U.S. National Oceanic and Atmospheric Administration. National Ocean Service. *Assessing the Social Costs of Oil Spills: The Amoco Cadiz Case Study.* Washington, DC: GPO, 1983. 144p.

Environmental Aspects

Atwood, D. K. "Petroleum Pollution in the Caribbean." *Oceanus* 30 (Winter 1987/1988): 25–32.

Frey, R. W., and others. "Coastal Sequences, Eastern Buzzards Bay, Massachusetts: Negligible Record of an Oil Spill." *Geology* 17 (May 1989): 461–465.

Jackson, J. B. C., and others. "Ecological Effects of a Major Oil Spill on Panamanian Coastal Marine Communities." *Science* 243 (January 6, 1989): 37–44.

Kiechel, Walter, III. "'The Admiralty Case of the Century': A Year after the Immense Oil Spill off the Coast of France, the Wrecked Amoco Cadiz Is Still Sending Forth Shock Waves: The Damages Claimed Run to Nearly $2 Billion." *Fortune* 99 (April 23, 1979): 78–82.

Lewis, Thomas A. "Tragedy in Alaska." *National Wildlife* 27 (June–July 1989): 4–9.

Lindén, O., and others. "Effects of Oil and Oil Dispersant on an Enclosed Marine Ecosystem." *Environmental Science and Technology* 21 (April 1987): 374–382.

Lindstedt-Siva, J. "Advance Planning for Dispersant Use: Minimizing the Impacts of Oil Spills." *ASTM Standardization News* 15 (April 1987): 36–40.

McFarland, W. "Air Stripping Removes Petroleum from Groundwater." *Water/Engineering & Management* 136 (May 1989): 48–52.

Neff, J. M. "Biological Effects of Oil in the Marine Environment." *Chemical Engineering Progress* 83 (November 1987): 27–33.

Nichols, A. B. "Oil Accident Ignites Response Debate." *Journal of Water Pollution Control Federation* 60 (April 1988): 466–472.

Pratt, Joseph A. "Growth or a Clean Environment? Response to Petroleum-Related Pollution in the Gulf Coast Refining Region [from Port Arthur, TX, to the Houston ship channel]." *Business History Review* 52 (Spring 1978): 1–29.

Shen, H. T., and P. D. Yapa. "Oil Slick Transport in Rivers." *Journal of Hydraulic Engineering* 114 (May 1988): 529–543.

"Special Report: Troubled Waters." *Amicus Journal* 11 (Summer 1989): 10–31.

Stiver, W., and D. Mackay. "Evaporation Rate of Spills of Hydrocarbons and Petroleum Mixtures." *Environmental Science and Technology* 18 (November 1984): 834–840.

Laws and Regulations

Bederman, David J. "High Stakes in the High Arctic: Jurisdiction and Compensation for Oil Pollution from Offshore Operations in the Beaufort Sea." *Alaska Law Review* 4 (June 1987): 37–69.

Dempsey, Paul Stephen. "Compliance and Enforcement in International Law: Oil Pollution of the Marine Environment by Ocean Vessels." *Northwestern Journal of International Law and Business* 6 (Summer 1984): 459–561.

Eaton, Judith R. "Oil Spill Liability and Compensation: Time To Clean Up the Law." *George Washington Journal of International Law and Economics* 19:3 (1985): 787–827.

Gaskell, N. J. J. "The Amoco Cadiz: Liability Issues." *Journal of Energy and Natural Resources Law* 3:3 (1985): 169–194; 3:4 (1985): 225–242.

Hartje, Volkmar J. "Oil Pollution Caused by Tanker Accidents: Liability versus Regulation." *Natural Resources Journal* 24 (January 1984): 41–60.

Jacobsen, Douglas A., and James D. Yellen. "Oil Pollution: The 1984 London Protocols and the Amoco Cadiz [oil tanker spill, March 1978; international pollution liability schemes]." *Journal of Maritime Law and Commerce* 15 (October 1984): 467–488.

Smets, Henri. "The Oil Spill Risk: Economic Assessment and Compensation Limit." *Journal of Maritime Law and Commerce* 14 (January 1983): 23–43.

Springall, R. C. "P & I Insurance and Oil Pollution." *Journal of Energy and Natural Resources Law* 6:1 (1988): 25–40.

"Superfund's Petroleum Exclusion." *Environmental Forum* 3 (August 1984): 32–38.

U.S. Congress. House. Committee on Merchant Marine and Fisheries. Subcommittee on Coast Guard and Navigation. *Oil Pollution Liability: Hearing, April 20, 1983, on H.R.2222 (H.R.2115, H.R.2368, a Bill To Provide a Comprehensive System of Liability and Compensation for Oil Spill Damage and Removal Costs, and for Other Purposes.* 98th Cong., 1st sess. Washington, DC: GPO, 1983. 414p.

————. *Oil Pollution Liability: Hearing, March 27, 1985, on H.R.1232, a Bill To Provide a Comprehensive System of Liability and Compensation for Oil Spill Damage and Removal Costs, and for Other Purposes.* 99th Cong., 1st sess. Washington, DC: GPO, 1985. 290p.

————. *Oil Pollution Liability: Hearing, March 31, 1987, on H.R.1632, a Bill To Provide a Comprehensive System of Liability and Compensation for Oil Spill Damage and Removal Costs, and for Other Purposes.* 100th Cong., 1st sess. Washington, DC: GPO, 1987. 203p.

U.S. Congress. House. Committee on Merchant Marine and Fisheries. Subcommittee on Fisheries and Wildlife Conservation and the Environment. *Review of Current Laws for Recovering Damages Caused by Spills of Oil and Hazardous Substances: Hearing, May 25, 1989, on Enforcement of Current Laws in Obtaining Compensation for Damages to Natural Resources Caused by Pollution, and Consideration of Whether New Legislation Is Needed.* 101st Cong., 1st sess. Washington, DC: GPO, 1989. 107p.

U.S. Congress. House. Committee on Public Works and Transportation. Subcommittee on Water Resources. *Oil Spill Liability and Compensation: Hearing, June 24, 1987.* 100th Cong., 1st sess. Washington, DC: GPO, 1987. 320p.

Asbestos

Chesson, J., and others. "Airborne Asbestos in Public Buildings." *Environmental Research* 51 (February 1990): 100–107.

Krizan, W. G. "Asbestos: Hazard and Hysteria." *ENR* 220 (June 2, 1988): 20–24.

Mossman, B. T., and others. "Asbestos: Scientific Developments and Implications for Public Policy." *Science* 247 (January 19, 1990): 294–301.

Pitt, R. "Asbestos as an Urban Area Pollutant." *Journal of Water Pollution Control Federation* 60 (November 1988): 1993–2001.

Sharplin, A. "Manville Lives On as Victims Continue To Die." *Business and Society Review* no. 65 (Spring 1988): 25–29.

Spurny, K. R. "On the Release of Asbestos Fibers from Weathered and Corroded Asbestos Cement Products." *Environmental Research* 48 (February 1989): 100–116.

U.S. Congress. House. Committee on Public Works and Transportation. Subcommittee on Public Buildings and Grounds. *Potential Health Hazards Associated with the Use of Asbestos-Containing Material in Public and Private Facilities: Hearings, February 24 and March 21, 1984.* 98th Cong., 2d sess. Washington, DC: GPO, 1984. 315p.

U.S. Congress. Senate. Committee on Environment and Public Works. Subcommittee on Toxic Substances and Environmental Oversight. *Hazardous Asbestos Assessment: Hearing, May 15, 1986, on S.2083 and S.2300, Bills Providing for the Abatement of Hazardous Asbestos.* 99th Cong., 2d sess. Washington, DC: GPO, 1986. 327p.

White, M. K., and others. "Physiological and Subjective Responses to Working in Disposable Protective Coveralls and Respirators Commonly Used by the Asbestos Abatement Industry." *American Industrial Hygiene Association Journal* 50 (June 1989): 313–319.

Selected Journal Titles

The journals listed below publish articles on many aspects of hazardous material and toxic waste. Because the environmental problems associated with these substances have become important only in recent years, new journals are continually appearing. For new journals and additional information, please consult *Ulrich's International Periodicals Directory, 1988–1989,* 28th edition (New York: R. R. Bowker Company, 1989, 3 vols.).

Sample Entry

Journal Title

1. Editor
2. Year first published
3. Frequency of publication
4. Code
5. Special features
6. Address of publisher

Journals

Amicus Journal

1. Peter Borrelli
2. 1979
3. Quarterly
4. ISSN 0276-7201
5. Bk. rev., illus., index, cum. index
6. National Resources Defense Council, Inc.
   122 East 42nd Street, Room 4500
   New York, NY 10168

Analytical Chemistry

1. George Morrison
2. 1929
3. Twice monthly
4. ISSN 0003-2700
5. Adv., bk. rev., abstr., bibl., charts, illus., tr. lit., index
6. American Chemical Society
   1155 16th Street, NW
   Washington, DC 20036

Appraisal Journal

1. Patricia Hollopeter
2. 1932
3. Quarterly
4. ISSN 0003-7087
5. Bk. rev., abstr., charts, illus., stat., index

6. American Institute of Real Estate Appraisers
National Association of Realtors
430 North Michigan Avenue
Chicago, IL 60611-4088

Ecology Law Quarterly

1. Richard Allen
2. 1971
3. Quarterly
4. ISSN 0046-1121
5. Adv., bk. rev., bibl., index
6. University of California Press
Journals Division
2120 Berkeley Way
Berkeley, CA 94720

Environmental Action

1. Editorial Board
2. 1970
3. Bimonthly
4. ISSN 0013-922X
5. Adv., bk. rev., film rev., illus., index
6. Environmental Action Inc.
1525 New Hampshire Avenue, NW
Washington, DC 20036

Environmental Law Reporter

1. Barry Breen
2. 1971
3. Monthly
4. ISSN 0046-2284
5. Bk. rev., index
6. Environmental Law Institute
1616 P Street, NW, Suite 200
Washington, DC 20036

Environmental Protection Agency Journal

1. John Heritage
2. 1975
3. Quarterly
4. ———
5. Adv.
6. U.S. Environmental Protection Agency
Office of Public Affairs

Waterside Mall
401 M Street, SW
Washington, DC 20460

Environmental Research

1. I. J. Selikoff
2. 1967
3. Bimonthly
4. ISSN 0013-9351
5. Adv., bk. rev., illus., index
6. Academic Press, Inc.
   Journal Division
   1250 Sixth Avenue
   San Diego, CA 92101

Environmental Science and Technology

1. Russell F. Christman
2. 1967
3. Monthly
4. ISSN 0013-936X
5. Adv., bk. rev., abstr., bibl., charts, illus., stat., tr. lit., index
6. American Chemical Society
   1155 16th Street, NW
   Washington, DC 20036

Harvard Environmental Law Review

1. ———
2. 1976
3. 2 times a year
4. ISSN 0147-8257
5. ———
6. Harvard University Law School
   Publications Center
   Cambridge, MA 02138

Journal of Environmental Engineering

1. ———
2. 1956
3. Bimonthly
4. ISSN 0733-9372
5. ———
6. American Society of Civil Engineers
   345 East 47th Street
   New York, NY 10017

Journal of Hazardous Materials

1. G. F. Bennett and R. F. Griffiths
2. 1975
3. 6 times a year
4. NE ISSN 0304-3894
5. Adv., bk. rev., index
6. Elsevier Science Publishers B.V.
   P.O. Box 211
   1000 A. E. Amsterdam
   Netherlands

Journal of the Air Pollution Control Association

1. Harold M. Englund
2. 1951
3. Monthly
4. ISSN 0002-2470
5. Adv., bk. rev., abstr., bibl., charts, illus., stat., index
6. Air Pollution Control Association
   Box 2861
   Pittsburgh, PA 15230

Management of World Wastes

1. Bill Wolpin
2. 1958
3. Monthly
4. ISSN 0745-6921
5. Adv., bk. rev., charts, illus., stat., tr. lit.
6. Communication Channels, Inc.
   6255 Barfield Road
   Atlanta, GA 30328

Natural Resources Journal

1. Albert E. Utton
2. 1961
3. Quarterly
4. ISSN 0028-0739
5. Adv., bk. rev., charts, index,
   cum. index every ten years
6. University of New Mexico
   School of Law
   1117 Stanford, NE
   Albuquerque, NM 87131

Nuclear and Chemical Waste Management

1. A. Moghissi
2. 1980
3. Quarterly
4. ISSN 0191-815X
5. Adv., illus.
6. Pergamon Press, Inc.
   Journals Division
   Maxwell House, Fairview Road
   Elmsford, NY 10523

Ocean Development and International Law

1. Daniel S. Cheever
2. 1973
3. Bimonthly
4. ISSN 0090-8320
5. Adv., bk. rev., abstr., charts, stat., index, cum. index
6. Taylor & Francis New York
   3 East 44th Street
   New York, NY 10017

Toxic Substances Journal

1. George S. Dominguez
2. 1979
3. Quarterly
4. ISSN 0199-3178
5. ———
6. Hemisphere Publishing Corporation
   79 Madison Avenue, Suite 1110
   New York, NY 10016-7892

Water Pollution Control Federation Journal

1. Peter J. Piecuch
2. 1928
3. Monthly
4. ISSN 0043-1303
5. Adv., bk. rev., illus., index, cum. index (vols. 1–42, 1928–1970)
6. Water Pollution Control Federation
   601 Wythe Street
   Alexandria, VA 22314-1994

6

Films, Filmstrips, and Videocassettes

A SELECTED NUMBER OF FILMS, filmstrips, and videocassettes present some aspects of the problems associated with hazardous material and toxic waste. Graphic presentations often convey environmental problems more vividly than can the written word. Because critical environmental problems caused by hazardous material and toxic waste may become evident only after an extended period of time, the general public may ignore the problems until a disastrous situation occurs. As these problems are identified, films provide visual evidence of environmental damage.

The following sources list films and videos in English:

AAAS Science Film Catalog. Washington, DC: American Association for the Advancement of Science, 1975. 398p.

Educational Film/Video Locator of the Consortium of University Film Center and R. R. Bowker. 4th ed. New York: R. R. Bowker Company, 1990–1991. 2 vols. 3,361p.

Film & Video Finder. 2d ed. Medford, NJ: Plexus Publishing Company, 1989. 3 vols. 1,424p.

Films in the Sciences: Reviews and Recommendations. Washington, DC: American Association for the Advancement of Science, 1980. 172p.

Index to Environmental Studies Multimedia. University Park, Los Angeles, CA: National Information Center for Educational Media (NICEM), University of Southern California, 1977. 1,113p.

Video Rating Guide for Libraries. Santa Barbara, CA: ABC-CLIO, quarterly.

The Video Source Book. 11th ed. Syosset, NY: National Video Clearing House, Inc., 1980. 2 vols. and supplements. 2,371p.

General

The Challenge of Survival Series
Walt Disney Educational Media Company
500 South Buena Vista Street
Burbank, CA 91521
Color, 11 minutes, Beta, VHS, ¾ " U-matic, 1987.

Examines use of land, water, and chemicals to make us understand the need for proper management of our resources. "The Challenge of Survival: Chemicals" is one of the main parts of the film.

Ecology: Barry Commoner's Viewpoint
Encyclopedia Britannica Educational Corp.
425 North Michigan Avenue
Chicago, IL 60611
Color, 19 minutes, sound, 16mm, 1977.

Presents Barry Commoner's philosophy of ecology. Shows humans' impact on the environment by use of chemicals.

Great Lakes: Troubled Waters
Umbrella Films
60 Blake Road
Brookline, MA 02146
Color, 53 minutes, sound, video, 1987.

Emphasizes the problem of toxic wastes and the threat they pose to the environment. Political attempts to block endeavors to find solutions to pollution problems are shown, as are scientists doing fieldwork to find solutions.

Hazardous Chemicals
Film Communicators
Distributed by
MIT Film and Video
108 Wilmot Road
Deerfield, IL 60015
Color, 12 minutes, sound, ¾ " or ½" videocassette, n.d.

Creator of Murphy's law describes how to handle hazardous chemicals, cleaning products, and solvents safely.

Hazardous Chemicals
The Film Library
626 South Westmoreland Avenue
Los Angeles, CA 90005
Color, 12 minutes, sound, VHS, ¾" U-matic, 1980s.

Examines solvents, chemicals, and cleaning products most frequently used in industry.

Hazardous Chemicals: Handle with Care
Educational Images
P.O. Box 3456, West Side
Elmira, NY 14905
Color, 60 minutes, sound, Beta, VHS, ¾" U-matic, 1984.

Shows the importance of hazardous chemicals in our lives but also the environmental problems caused by their use and disposal.

Hazardous Materials—Handle with Care
The Film Library
626 South Westmoreland Avenue
Los Angeles, CA 90005
Color, 13 minutes, sound, VHS, ¾" U-matic, 1980s.

Looks at how to handle solvents and corrosives in industry.

Hazardous Toxic Materials
Film Communicators
Distributed by
MIT Film and Video
108 Wilmot Road
Deerfield, IL 60015
Color, 30 minutes, sound, ¾" or ½" videocassette, n.d.

Provides an introduction to the wide variety of toxic materials in the environment.

Toxic Earth—The Need To Unite
AFL-CIO
Educational Department
815 Sixteenth Street, NW
Washington, DC 20006
Color, 18 minutes, sound, ½" videocassette, Beta/VHS.

Describes the growing problem of toxic chemicals in the environment and the need to formulate plans for their removal.

Toxic Trials
WGBH-TV
125 Western Avenue
Boston, MA 02134
Color, 58 minutes, sound, ½" VHS, 1985.

Investigates the controversy surrounding a lawsuit against a local chemical company brought by an American community affected by hazardous waste. Examines the scientific, medical, and legal expertise amassed by both sides in this landmark case, and assesses its potential legal and scientific ramifications for future cases.

Toxic Waste
Downtown Community TV Center
Wyoming University
Cheyenne, WY 82003
Color, 25 minutes, sound, ½" VHS, 1984.

Documentary examines how people's lives are affected when they become victims of toxic waste and how one community banded together to close down one of the largest toxic waste handlers in the country. Reveals much of the activity of handling toxic waste in the early years of the Reagan administration.

Toxic Waste in America
Downtown Community Television Center
87 Lafayette Street
New York, NY 10013
Color, 25 minutes, sound, ½" reel, ¾" U-matic, format other than listed available, 1984.

Examines how people's lives can be destroyed through toxic waste.

A Toxicologist
Hawkhill Associates Inc.
125 East Gilman Street
Madison, WI 53703
Color, 10 minutes, sound, video, 1 filmstrip and 1 cassette, 1986.

Filmstrip is from *People in Science Today* and gives an interview with Colin Jefcoate, head of a toxicology research laboratory at the University of Wisconsin. He explains the study of the interaction of chemicals with living systems involving people with various backgrounds. Toxicology is a young science. The dumping of wastes and the controlling of toxic residues are examined.

Water Pollution
Sterling Educational Films
Division of Walter Reade Organization, Inc.
241 East 34th Street
New York, NY 10016
Color, 10 minutes, sound, 16mm, n.d.

Graphically traces and illustrates the many causes of water pollution. As garbage, sewage, insecticides, and industrial waste flow into our rivers and lakes, clean water becomes contaminated and useless.

Water Pollution: A First Film
BFA Educational Media
2211 Michigan Avenue
Santa Monica, CA 90404
Color, 8 minutes, sound, 16mm, n.d.

Follows a stream from its beginning to its end, showing the many sources of pollution.

Management

Haz Mat Emergency
Emergency Film Group
1380 Soldiers Field Road
Boston, MA 02135
Color, 160 minutes, sound, Beta, VHS, ¾" U-matic, 1988.

A series of programs providing information on how to deal with hazardous material, including recognizing hazards, wearing protective clothing, and conducting safe operations.

Hazardous Area Identification
MTI Teleprograms/Coronet International
Distributed by
Simon and Schuster
108 Wilmot Avenue
Deerfield, IL 60015
Color, 9 minutes, sound, ¾" or ½" videocassette, n.d.

Shows how to recognize hazardous areas and take protective measures, including barricading, taping, and tagging.

Hazardous Cargo—Accepting and Airlifting
U.S. National Audiovisual Center
8700 Edgeworth Drive
Capitol Heights, MD 20743-3701
Color, 28 minutes, sound, 16mm film, ¾" or ½" video, 1977.

Discusses the importance of properly preparing hazardous materials for shipment by air, and emphasizes dangers of careless storage. Depicts marking, identifying, and inspecting hazardous cargo.

Hazardous Materials—Emergency Response
Courter Films and Assoc.
121 Northwest Crystal Street
Crystal River, FL 32629
Color, 31 minutes, sound, 16mm film, optical sound, 1979.

Shows how to respond to and handle emergencies involving hazardous materials.

Hazardous Waste Management
Gulf Publishing Company Video
P.O. Box 2608
Houston, TX 77001
Color, 20 minutes, sound, ¾" or ½" videocassette, 1986.

Fits into company training efforts concerning hazardous waste products as in the Resource Conservation and Recovery Act. Gives legal and environmental concerns of handling and shows who has what responsibilities.

Hazardous Waste Management
International Training Company
3301 Allen Parkway
P.O. Box 3881
Houston, TX 77001
Color, 20 minutes, sound, Beta, VHS, ¾" U-matic, 1984.

Purpose of film is to inform and motivate industrial personnel to work as a team in managing hazardous material.

Poison and the Pentagon
PBS Video
1320 Braddock Place
Alexandria, VA 22314-1698
Color, 58 minutes, sound, Beta, VHS, ¾" U-matic, 1988.

Investigates the U.S. government's negligence in cleaning up toxic wastes.

Toxic Chemicals
Bullfrog Films Inc.
Oley, PA 19547
Color, 52 minutes, sound, Beta, VHS, ¾" U-matic, 2 progs., 1986.

Shows why a plan for chemical emergencies is important for local community officials and the general public to know.

Toxic and Hazardous Waste Disposal

Hazardous Waste
Direct Cinema Limited Inc.
P.O. Box 69589
Los Angeles, CA 90069
Color, 35 minutes, sound, Beta, VHS, ¾" U-matic, format other than listed available, 1984.

Documentary narrated by Hal Holbrook tells how concerned citizens organized to clean up some of the 17,000 toxic chemical dumps littering the countryside.

Hazardous Waste
Emergency Film Group
1380 Soldiers Field Road
Boston, MA 02135
Color, 60 minutes, sound, Beta, VHS, ¾" U-matic, 1989.

Considers measures to deal with hazardous waste in emergency spills.

Hazardous Waste—The Search for Solutions
AFL-CIO
Educational Department
815 Sixteenth Street, NW
Washington, DC 20006
Color, 30 minutes, sound, 16mm film, optical sound, 1984.

Depicts different approaches that citizens of five states have developed to protect themselves from the effects of toxic chemical dumps.

Hazardous Waste: Who Bears the Cost?
Indiana University
Audio-Visual Center
Bloomington, IN 47405
Color, 28 minutes, sound, 16mm, 1981.

Examines the problem of the safe management of hazardous waste through a presentation of the experiences of the citizens of Woburn, Massachusetts, where the leather and chemical industries of the nineteenth century left many sites where hazardous waste had been dumped. It was feared that these dumps had polluted the groundwater and caused increased rates of cancer. The solution of the problem involved the development of treatment facilities to deal with the dumping of the past.

Hazardous Waste Options
Stuart Finley, Inc.
3428 Mansfield Road
Falls Church, VA 22041
Color, 22 minutes, sound, 16mm film, optical sound, 1981.

Describes how technology has provided appropriate methods of handling hazardous wastes—including waste treatment and conversion, secure land-fills, high-temperature incineration, deep well injection, and land treatment—that can result in real environmental danger if mismanaged.

The Hazcom Reports
U.S. Training Inc.
11246 South Post Oak, Suite 400
Houston, TX 77035
Color, 15 minutes, sound, Beta, VHS, ¾″ U-matic, 5 progs., 1987.

Series of five tapes telling workers how to handle dangerous chemicals safely according to OSHA standards.

Landfill
Stuart Finley, Inc.
3428 Mansfield Road
Falls Church, VA 22041
Color, 12 minutes, sound, 16mm, n.d.

Proves that hazardous waste disposal need not be damaging to the environ-ment. Depicts a regional transfer and landfill operation that serves a big city and suburbs.

A Love Canal Family
Pennsylvania State University
Audio Visual Aids Library
University Park, PA 16802
Color, 28 minutes, sound, V-U ¾″ U-matic, 1980.

Story of a family who discovered that their Love Canal neighborhood in Niagara Falls, New York, was sitting on top of one of the nation's largest chemical dumping grounds. The death of their son shattered their lives.

Documents the case and the radicalization of parents who believe their son died from exposure to chemical waste.

The Toxic Goldrush
WNET-TV
356 West 58th Street
New York, NY 10019
Color, 30 minutes, color, ¾" videocassette, Beta/VHS.

Describes the great profits to companies involved in toxic waste cleanup.

Toxic Waste: Information Is the Best Defense
Bullfrog Films Inc.
Oley, PA 19547
Color, 26 minutes, sound, Beta, VHS, ¾" U-matic, 1986.

Film in two parts shows how community groups can organize protests and ordinances to fight toxic dumping.

Toxic Wastes
Hampton Roads Education TV Assn.
Hampton Roads, VA 23669
Color, 14 minutes, sound, ½" VHS, 1985.

Depicts the dumping of chemical-filled barrels in a remote rural region in the early 1950s. Although a warning is posted, the film illustrates how a typical forgotten toxic disposal site ultimately affects the environment.

Health

Can I Drink the Water?
University of California
Extension Media Center
2176 Shattuck Avenue
Berkeley, CA 94704
Color, 27 minutes, sound, video, 1987.

Examines drinking water—surface, ground, and bottled—in three areas of California. Discussed are salty water from the delta area, surface water polluted with agricultural pesticide (DBCP), and groundwater containing an industrial solvent (TCA) and nitrates. Viewpoints of industrialists, environmentalists, government officials, citizens, and so on are given. Some issues raised for discussion are the desire for 100 percent pure water, the effect of sodium on blood pressure, and the effect of toxic dumps on water quality.

Hazardous Chemical Risks: Employee Rights
RMI Media Productions Inc.
2807 West 47th Street
Shawnee Mission, KS 66205
Color, 120 minutes, sound, Beta, VHS, ¾" U-matic, 1987.

Tells about Iowa's recently enacted law requiring employers to inform employees about any hazardous chemicals where they work. Also includes a question-and-answer session.

Hazardous Materials—Flammables
The Film Library
626 South Westmoreland Avenue
Los Angeles, CA 90005
Color, 14 minutes, sound, VHS, ¾" U-matic, 1980s.

Stresses responsibility of employees to follow company procedure and safety rules when handling chemicals and flammables.

The Hazcom Reports
U.S. Training Inc.
11246 South Post Oak, Suite 400
Houston, TX 77035
Color, 15 minutes, sound, Beta, VHS, ¾" U-matic, 1987.

Explains to workers how to use dangerous chemicals safely according to OSHA standards.

It Can't Happen Here
Commonwealth Films Inc.
223 Commonwealth Avenue
Boston, MA 02116
Color, 39 minutes, sound, Beta, VHS, ¾" U-matic, 1987.

The purpose of the film is to train personnel to meet operational requirements of waste disposal and comply with hazardous and toxic environmental laws.

Toxic Hazards in Industry
The Film Library
616 South Westmoreland Avenue
Los Angeles, CA 90005
Color, 23 minutes, sound, VHS, ¾" U-matic, 1980s.

Shows how workers can protect themselves when working with toxic substances.

Toxics in the Workplace
New York State Education Department
Center for Learning Technologies, Media Distribution Network

Cultural Education Center, Room C-7, Concourse Level
Albany, NY 12230
Color, 30 minutes, sound, Beta, VHS, ½" reel, ¾" U-matic, 2" quad, 1984.

Shows an electrical worker fighting the Massachusetts Legislature about the right-to-know legislation calling for labeling of toxic chemicals in the workplace.

Pesticides

Agricultural Update: Pest Management
U.S. Department of Agriculture
Motion Picture Service
South Building, Room 1850
Washington, DC 20250
Color, 53 minutes, sound, 16mm, 1978.

Discusses the successes and problems of pest management programs of the USDA and the different approaches used in various states.

Budworks
Green Mountain Post Films Inc.
P.O. Box 229
Turner Falls, MA 01376
Color, 35 minutes, sound, Beta, VHS, ¾" U-matic, 1979.

Shows the conflict over the timberland spraying program in New Brunswick since 1952 that barely kills the insects but does harm the environment.

Pesticide Politics
UC Video
425 Ontario Street, SE
Minneapolis, MN 55414
Black and white, 23 minutes, sound, ¾" U-matic, format other than listed available, 1976.

Shows legal and civil actions against the Forestry Service to stop the spraying of 245-T taken by people in the Britt, Minnesota, area. The chemical caused health problems for people, animals, and vegetation.

Pesticides
Emergency Film Group
1380 Soldiers Field Road
Boston, MA 02135
Color, 60 minutes, sound, Beta, VHS, ¾" U-matic, 1989.

The purpose of this film is to help emergency medical professionals to be able to respond to problems involving pesticides.

Pesticides
Marshfield Regional Video Network
100 North Oak Avenue
Marshfield, WI 54449
Color, 60 minutes, sound, Beta, VHS, ¾″ U-matic, 1981.

Discusses the toxicity demographics and absorption of pesticides.

Pesticides
Q-Ed Productions
Division of Cathedral Films
P.O. Box 1608
Burbank, CA 91507
Color, 63 frames, sound filmstrip-audiotape, 1971.

Points out that one of the greatest threats to the earth's ecosystem is human use of pesticides, particularly DDT. Shows extensive damage to different life systems caused by DDT.

Pesticides—Fundamentals of Application
Farmland Industries
University of Missouri
Columbia, MO 65211
Color, 21 minutes, sound, 16mm, 1968.

Discusses the proper application of spraying chemicals on the land. Explains the different kinds of pumps and the chemicals that should be used and stresses the proper use of equipment.

Pesticides: The Hidden Assassins
Video Out
261 Powell Street
Vancouver, B.C.
Canada V6A 1G3
Color, 29 minutes, sound, Beta, VHS, ¾″ U-matic, 1979.

Examines use of pesticide spraying in British Columbia.

Pesticides and Pest Management
Plaid Productions—Personnel Training Programs
1645 Hicks Road
Rolling Meadows, IL 60008
Color, 26 minutes, sound, ½″ or ¾″ videocassette, n.d.

Discusses pest management problems, pesticides, and regulations controlling their use as applicable to supervisory monitoring of in-house pest activity and control.

Pesticides and Principles:
From "For Export Only"
Harvard Business School
Harvard University
Boston, MA 02163
Color, 26 minutes, sound, 1/2" VHS, 3/4" U-matic, 1985.

A report on the "circle of poison" that results when chemicals, sometimes illegal in the United States, are shipped to Third World countries. When these exported chemicals are misused, they can be reintroduced to the United States through residue in imports.

Pesticides and the Environment
NETCHE (Nebraska ETV Council for Higher Education)
Box 8311
Lincoln, NE 68501
Color, 30 minutes, sound, 1/2" open reel, 3/4" U-matic, 1975.

Film has three programs showing research that has been done to help us understand pesticides and their effects.

Pesticides and Wildlife
Tennessee Game and Fish Commission
Box 40707
Nashville, TN 37220
Color, 30 frames, script, n.d.

Describes several incidents of animal mortality following the application of chemical pesticides.

The Poisoned Planet
Contemporary McGraw-Hill Films
330 West 42nd Street
New York, NY 10036
Color, 19 minutes, sound, 16mm, n.d.

The question is asked, Can pesticides that are supposed to save our food supply also end our lives? DDT, mercury, and soft pesticides are discussed as to their effects on the environment.

The Poisoning of Michigan
Media Guild
11526 Sorrento Valley Road, Suite J
San Diego, CA 92121
Color, 65 minutes, sound, Beta, VHS, 3/4" U-matic, 1977.

Explains the effects of PCBs on the environment of Michigan.

The Rise and Fall of DDT

Time-Life Films
Time Life Building, 32nd Floor
Rockefeller Center
New York, NY 10020
Color, 18 minutes, sound, 16mm, 1976.

Reviews the usefulness of DDT in combating insects and shows the extreme charges made against the use of DDT, such as in the reproduction of birds. The film presents a particularly pessimistic point of view.

Oil Spills

The Great Oil Disaster

Coronet Films & Video
108 Wilmot Road
Deerfield, IL 60015
Color, 30 minutes, sound, 16mm, video, 1987.

Documentary on the *Amoco Cadiz* oil spill begins by showing how oil spills occur and shows cleanup efforts and the effects years later. Shows the mess an oil spill can cause, and concludes with ways to prevent future spills and new procedures for cleanup use.

Oil Spill Contingency Planning

NUS Training Corporation
910 Clopper Road
Gaithersburg, MD 20878-1399
Color, 30 minutes, sound, Beta, VHS, ½" reel, ¾" U-matic, 1978.

Series of eight training programs to show personnel how to prevent and minimize oil spills that occur during extraction, refining, distribution, and use of petroleum products. Programs available individually.

Oil Spill Response Training Program

NUS Training Corporation
910 Clopper Road
Gaithersburg, MD 20878-1399
Color, 30 minutes, sound, Beta, VHS, ½" reel, ¾" U-matic, 1978.

Film has 23 programs to train and indoctrinate personnel in preventing and minimizing oil spills that could occur during extraction, refining, distribution, and use of petroleum products. Programs available individually or as units.

Oil Spills
University of Toronto
Media Centre/Distribution Office
121 St. George Street
Toronto, Ontario
Canada M5S 1A1
Color, 7 minutes, sound, Beta, VHS, 3/4″ U-matic, 1977.

Deals with research being conducted on oil properties in

Glossary

abatement   The reduction in degree or intensity of hazardous waste.

acclimation   The physiological and behavioral adjustments of an organism to environmental changes.

acid   A compound that contains hydrogen that reacts with a base or metal to form a salt. The metal replaces the hydrogen.

activated sludge   Sludge that has been aerated and subjected to bacterial action. Used to increase rate of breakdown of organic matter in raw sewage during secondary waste treatment.

acute toxicity   Any poisonous effect produced within a short period of time, usually one to four days after contact with a toxic substance, resulting in sickness and often death.

aeration   Circulation of oxygen through a substance, as in wastewater treatment, where it aids in purification.

aquifer   An underground layer of bedrock that contains significant amounts of groundwater.

asbestos   A group of toxic minerals found in nature as compact, long, silky fibers that are fireproof and have good electrical and thermal-insulating properties. Can cause cancer if inhaled or ingested in air or water.

bacteria   Single-celled microorganisms. Some aid in pollution control by breaking down organic matter in air and water.

band application In pesticides, the spreading of chemicals over or next to each row of plants in a field.

basal application In pesticides, the spreading of a chemical on vegetation just above the soil line.

beryllium A metal that can be hazardous to human health when inhaled. It is a discharge in machine shops, ceramic factories, and propellant plants.

bill of lading A receipt listing goods shipped, issued by a common carrier.

biodegradation Any substance that is chemically decomposed quickly through the action of microorganisms.

cadmium A heavy metal element that accumulates in the environment.

carcinogen Any substance that produces or incites cancer.

catalyst A substance that changes the rate of a chemical reaction but is itself unchanged at the end of the reaction.

chemical change Any change in a body or substance that changes its chemical composition.

chemical contamination An uncontrolled release of hazardous material that can contaminate an area.

chlorination The application of chlorine to water, sewage, or industrial waste to disinfect or to oxidize hazardous waste.

comminution Mechanical shredding or pulverizing of waste, used in solid waste management.

compound A substance consisting of two or more elements chemically united in definite proportions by weight.

containment Method or technology that prevents migration of hazardous waste into the environment.

DDT The first chlorinated hydrocarbon insecticide. It has a half life of 15 years and can collect in the fatty tissues of certain animals.

dermal toxicity The ability of a pesticide or toxic chemical to poison animals and people by touching skin.

desiccant A chemical agent that dries out plants or insects, causing death.

dilution ratio The relationship between the volume of water in a stream and the volume of incoming waste. It is a factor in the ability of a stream to assimilate hazardous waste.

effluent Treated or untreated waste material discharged into the environment.

elements One of the 106 known substances that cannot be divided into simpler substances except by nuclear disintegration. All matter is composed of elements.

environment The totality of all external conditions affecting the life, development, and survival of the organism.

Environmental Impact Statement A document required of federal agencies by the National Environmental Policy Act for major projects or legislative proposals.

enzyme Any of various organic substances that are produced in plant and animal cells and cause changes in other substances by catalytic action.

epidemiology The study of diseases as they affect populations.

exposure People or environments that are or may be exposed to the harmful effects of a hazardous materials emergency.

floc A clump of solids formed in waste by biological or chemical action.

flocculation Separation of suspended solids during wastewater treatment by chemicals.

fogging Applying a pesticide by rapidly heating the liquid chemical so that it forms very fine droplets that resemble smoke.

fumigant A pesticide that is vaporized to kill pests.

germicide Any compound that kills disease-carrying microorganisms. Germicides must be registered by the EPA.

habitat The sum of environmental conditions in a specific place for an organism, population, or community.

Hazard Ranking System The model used to determine inclusion of a waste site on the U.S. Environmental Protection Agency's National Priorities List for Superfund cleanup.

hazardous material Material that poses environmental or health hazards.

hazardous waste Material with characteristics of ignitability, corrosivity, reactivity, and toxicity as defined by the Environmental Protection Agency.

heavy metal Any metal with a specific gravity four times as dense as water, such as mercury, chromium, cadmium, arsenic, and lead. Generally toxic to living organisms at low concentrations.

herbicide A chemical that controls or destroys undesirable plants.

hydrocarbon Compounds that contain only hydrogen and carbon.

ignitability One of the four characteristics of a hazardous waste, referring to a waste's capability to burn.

impermeable Substance resistant to the flow of liquids through it.

incineration A method of thermally treating hazardous waste by burning it under carefully controlled conditions.

lagoon A shallow pond where sunlight, bacterial action, and oxygen work to purify waste material.

landfill A disposal site for hazardous waste involving its shallow burial.

leachate Liquids that have been leached (i.e., passed out or through by percolation) from a landfill or other hazardous waste facility.

leaching The process by which nutrient chemicals or contaminants are dissolved and carried away by water.

mercury A heavy metal, highly toxic if breathed or swallowed. Accumulates in the environment.

midnight dumping The disposal of hazardous waste by illegal methods.

minimization Any method to reduce the volume of hazardous waste by reducing the volume of materials.

molecule Smallest particle of an element or compound existing independently while keeping the characteristics of the element or compound.

mutagen Any substance that causes changes in the genetic structure in subsequent generations.

neutralization Reaction between an acid and a base that destroys the distinctive or active properties of both, as an alkali neutralizes an acid.

oil spill An accidental discharge of oil in water bodies. Control is attempted through chemical dispersion, combustion, mechanical containment, and absorption.

open burning Uncontrolled fires in an open dump.

organophosphates Pesticide chemicals that contain phosphorous, used to control insects. Most are short-lived, but some are toxic.

pH A measure of the acidity or alkalinity of a material, solid, or liquid. pH is represented by a scale of 0 to 14, with 7 being a neutral state, 0 most acids, and 14 most alkalines.

ppm (parts per million) A concentration equivalent to .0001 percent that is used to measure chemical contamination.

permeable Substance that is capable of letting liquids flow through it.

persistent pesticides Pesticides that do not break down chemically and remain in the environment after a growing season.

pesticide Any substance used to control pests ranging from weeds and insects to algae and fungi.

pesticide tolerance The amount of pesticide residue allowed by law to remain in or on a harvested crop. EPA sets standards well below the points where the chemicals might be harmful to consumers.

physical change Any change in a body or substance that does not change the chemical composition.

pollutant Any substance that adversely affects the usefulness of a resource.

porous Having tiny holes through which liquid or gases may pass.

protective clothing Clothing that prevents hazardous materials such as gases, vapors, liquids, and solids from coming in contact with the skin.

refuse reclamation Conversion of solid waste into useful products.

salt A compound formed by the reaction of an acid and a base.

sanitary landfill Landfill designed to protect the environment and still allow disposal of solid wastes.

selective pesticide A chemical designed to affect only certain types of pests, leaving other plants and animals unharmed.

sludge A waste that appears semisolid and is often composed of 90 percent or more water.

slurry A water mixture of insoluble matter that results from some pollution control techniques.

solid waste All waste collectively, including garbage, refuse, and other discarded material.

solid waste management Supervised handling of waste materials from their source through recovery processes to disposal.

solvent A chemical, usually a liquid, capable of dissolving other substances in it.

sump A depression or tank that catches liquid runoff for drainage or disposal.

Superfund A fund created by the Comprehensive Environmental Response, Compensation and Liability Act of 1980. Pays for the cleanup and removal of released hazardous substances at abandoned hazardous waste sites. The fund is chiefly generated by taxing chemical industries.

surface impoundment A low place or pit enclosed so as to hold a liquid.

toxic waste Discarded pollutants consisting of chemical products harmful to living organisms.

toxicant A chemical that controls pests by killing rather than repelling.

toxicology The science of poisons, their effects, antidotes, and detection.

variance Government permission for a delay or exception in the application of a given law, ordinance, or regulation.

waste Unwanted materials left over from manufacturing processes.

water table Upper level of groundwater.

Index